W9-CXT-083

FIBER OPTICS

FIBER
OPTICS

EDWARD A. LACY

63148

PRENTICE-HALL, INC., Englewood Cliffs, New Jersey 07632

Library of Congress Cataloging in Publication Data

Lacy, Edward A., date
  Fiber optics.

  Bibliography: p.
  Includes index.
  1. Optical communications.   2. Fiber optics.
  I. Title.
  TK5103.59.L3   621.38'0414   81-14374
  ISBN 0-13-314278-7   AACR2

Editorial/production supervision by Lori Opre
Cover design by Edsal Enterprises
Manufacturing buyer: Gordon Osbourne

Printed in the United States of America

10   9   8   7   6   5   4   3   2   1

ISBN   0-13-314278-7

PRENTICE-HALL INTERNATIONAL, INC., *London*
PRENTICE-HALL OF AUSTRALIA PTY. LIMITED, *Sydney*
PRENTICE-HALL OF CANADA, LTD., *Toronto*
PRENTICE-HALL OF INDIA PRIVATE LIMITED, *New Delhi*
PRENTICE-HALL OF JAPAN, INC., *Tokyo*
PRENTICE-HALL OF SOUTHEAST ASIA PTE. LTD., *Singapore*
WHITEHALL BOOKS LIMITED, *Wellington, New Zealand*

Dedicated to Mark, Valerie, and Kendra Lacy

CONTENTS

2 FUNDAMENTALS OF LIGHT   22

3 LIGHT SOURCES AND TRANSMITTERS   39

4 OPTICAL FIBERS   69

9 INTEGRATED OPTOELECTRONICS   181

PREFACE

Throughout the world fiber optics is now being used to transmit voice, television, and data signals by light waves over flexible hair-thin threads of glass or plastic. In such use it has significant advantages, as compared to conventional coaxial cable or twisted wire pairs. Consequently, millions of dollars are being spent to put these light-wave communication systems into operation.

To operate and maintain fiber-optic systems, present-day telecommunications technicians will be forced to learn the basics of fiber-optic components and systems.

This book has been prepared to give the average electronics technician the practical foundation for this challenging innovation. From the introduction to the appendixes, this book is arranged in a logical fashion to describe components and systems. No prior knowledge of optics is necessary.

This text covers the communications use of fiber optics. Other uses—such as medical and industrial inspection of inaccessible areas—are not discussed.

Edward A. Lacy

Satellite Beach, Florida

PREFACE

ACKNOWLEDGMENTS

It is with gratitude that I acknowledge the assistance of the following companies, individuals, organizations, and publishers in the preparation of this book:

AEG-Telefunken
AMP
Amphenol
Belden Corporation
Bell Canada
Bell Laboratories RECORD
Bell Telephone Laboratories
Cablesystems Engineering
Corning
Du Pont
EDN (Cahners Publishing Co.)
Electrical Communication
Electronic Design (Hayden Publishing Co., Inc.)
Electro-Optical Systems Design (Milton S. Kiver Publications, Inc.)
Fiberguide
Fujitsu
General Cable Company
General Optronics Corporation

GTE Lenkurt
Hamamatsu
Harris Corporation
Hewlett-Packard
Institute of Electrical and Electronics Engineers (IEEE)
International Business Machines (IBM)
International Telephone and Telegraph (ITT)
ITT Cannon
John Wiley & Sons Ltd.
Laser Focus (Advanced Technology Publications)
Machine Design
Math Associates
Mechanical Engineering
Motorola Inc.
National Communication System
Optical Information Systems (Exxon Enterprises, Inc.)
Photodyne
Prentice-Hall, Inc.
Professor Morris Grossman
Quartz & Silice
RCA Corporation
Siecor
Siemens AG
T&B/Ansley
Telephone Engineer and Management
Times Wire & Cable Co.
TRW Cinch

Particular thanks to AMP Incorporated for their permission to use numerous drawings, quotes, and paraphrases from their excellent publication, "Introduction to Fiber Optics and AMP Fiber-Optic Products," HB 5444.

And thanks to my wife Rita for her support in ways too numerous to mention.

E.A.L.

FIBER OPTICS

1

FIBER OPTICS: LIGHT-WAVE COMMUNICATIONS

1.1 INTRODUCTION

For years fiber optics has been merely a system for piping light around corners and into inaccessible places so as to allow the hidden to be seen. But now fiber optics has evolved into a system of significantly greater importance and use. Throughout the world it is now being used to transmit voice, television, and data signals by light waves over flexible hair-thin threads of glass or plastic. Its advantages in such use, as compared to conventional coaxial cable or twisted wire pairs, are fantastic. As a result, millions of dollars are being spent to put these light-wave communication systems into operation.

No longer a mere laboratory curiosity, fiber optics is now an important new proven technology, a recognized reality. In fact, some are calling it an exciting revolution which may affect our lives as much as computers and integrated circuits have. In the world of communications fiber optics is being compared in importance with microwave and satellite transmission.

It is indeed a new era in communications, the age of optical communications. In many ways it is a radical departure from the electronic communications we have become so accustomed to. Now instead of electrons moving back and forth over metallic wires to carry our signals, light waves are being guided by tiny fibers of glass or plastic to accomplish the same purpose.

With a bandwidth or information capacity thousands of times greater than that of copper circuits, fiber optics may soon provide us with all the communication paths we could ever want, at a price we can afford. It will make practical such services as two-way television that were too costly before the

1

development of fiber optics. In addition to a tremendous bandwidth, fiber optics has smaller and lighter cables than do copper-conductor systems, immunity to electrical noise, and numerous other advantages.

It is no wonder, then, that fiber optics is having a major impact on the electronics industry. Numerous companies, both new and old, large and small, are getting into the act, together with government agencies and military services. Thousands of engineers and scientists around the world are now involved in research and development of fiber-optic components and systems. Hundreds of technical papers are being presented as these people make technological breakthroughs.

Dozens of systems are now in operation and many others are being planned and installed. Field tests have shown that these systems can, without a doubt, meet their impressive claims. Off-the-shelf equipment is now available. As a result, current sales of $90 million a year [1] are expected to expand to $10 billion yearly by the year 2000 [2].

To date, fiber optics has found its greatest application in the telephone industry. But its other applications for transmitting data are vitally important in many other areas, such as computers, cable television, and industrial instrumentation. Still other uses are expected to be found as the price of fiber-optic systems drops and components are perfected.

Because of the advantages of fiber optics, some designers believe that any new communication system that does not use fiber optics, or at least consider its use, is obsolete even before it has been built. Although this may not always be the case with communication systems, it is becoming more and more obvious that the average technician may also become obsolete if he or she fails to master the basics of this new technology. After all, it will be up to the technician, not the engineer, to repair and maintain fiber-optic systems wherever they are used.

1.2 BASIC THEORY OF OPERATION

Fiber optics can be defined as that branch of optics which deals with communication by transmission of light through ultrapure fibers of glass or plastic. It has become the mainstay or major interest in the world of electro-optics, the blending of the technology of optics and electronics.

In a fiber-optic system or link (see Fig. 1.1) three major parts perform this task of communication: a light source, an optical fiber, and a light detector or receiver. The *light source* can be either a light-emitting diode (LED) or a semiconductor laser diode. The *optical fiber* can be a strand as short as 1 m or as long as 7 km. The *detector* can be either an avalanche photodiode (APD) or a positive-intrinsic-negative (PIN) diode. Each of these devices is discussed in detail in later chapters. For now, however, we want to see how they can be combined to form a communication system.

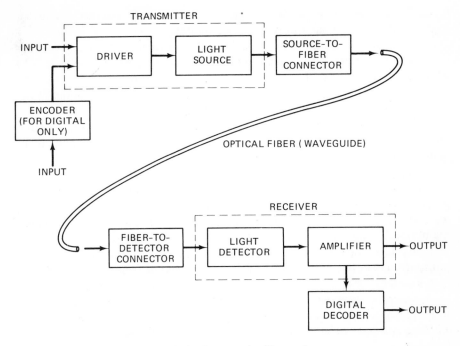

Figure 1.1   Basic elements of a fiber-optic system.

Basically, a fiber-optic system simply converts an electrical signal to an infrared light signal, launches or transmits this light signal onto an optical fiber, and then captures the signal on the other end, where it reconverts it to an electrical signal.

Two types of light-wave modulation are possible: analog and digital. In *analog modulation* the intensity of the light beam from the laser or LED is varied continuously. That is, the light source emits a continuous beam of varying intensity.

In *digital modulation,* on the other hand, the intensity is changed impulsively, in an on/off fashion. The light flashes on and off at an extremely fast rate. In the most typical system—pulse-code modulation (PCM)—the analog input signals are sampled for wave height. For voice signals this is usually at a rate of 8000 times a second. Each wave height is then assigned an 8-bit binary number which is transmitted in a series of individual time slots or slices to the light source. In transmitting this binary number, a 1 can be represented as a pulse of light and a 0 by the absence of light in a specific time slice.

Digital modulation is far more popular, as it allows greater transmission distances with the same power than analog modulation. Analog modulation is simpler, however, as shown in Fig. 1.1.

Notice that the encoder and decoder circuits are not necessary for analog modulation. The driver converts the incoming signal, whether digital or analog, into a form that will operate the source.

Even though miniature or tiny light sources and detectors are in use, optical fibers are so small that special connectors must be used to couple the light from the source to the fiber and from the fiber to the detector.

Not shown in Fig. 1.1 are miscellaneous connectors for the optical fiber which allow easier installation and disassembly for repair.

The optical fiber provides a low-loss path for the light to follow from the light source to the light detector. In a sense it is a waveguide that carries optical energy.

Most often this fiber is made of ultrapure glass, although plastic fibers are useful in a few applications. The glass fibers are so pure that they make eyeglass lenses seem opaque in comparison, according to Western Electric engineers. If a window of this material was made 1 km thick, say Bell Canada engineers, it would be as transparent as an ordinary pane of glass.

In many situations, of course, glass is considered to be a hard and brittle substance. However, optical fiber made of glass can be bent (see Fig. 1.2) and even knotted, yet it is stronger than stainless steel wire of the same diameter.

An optical fiber used for telecommunications typically consists of a glass core approximately 5 one-thousandths of an inch in diameter. Surrounding this core is a layer of glass or plastic, called a *cladding,* which keeps the light

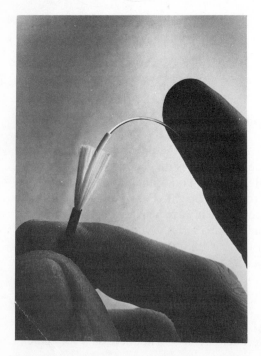

Figure 1.2 Optical fibers can be bent. (Courtesy of DuPont.)

waves within the core. Polyurethane jackets are added to the fiber to protect it from abrasion, crushing, chemicals, and the environment.

Individual fibers are often grouped to form cables. A typical fiber-optic cable may contain 1 to 24 of these fibers plus in some cases a steel wire that adds strength to the cable. The entire cable is still much smaller in diameter and much lighter than a comparable copper cable.

Although two-way or bidirectional fiber systems are being developed, for now most fibers are strictly one-way devices. Thus, two fibers working in pairs are needed for telephone conversations: one strand to transmit a voice from one end of the link, while the other carries the voice from the opposite end.

At the end of the fiber, the light signals are coupled to either an avalanche photodiode (APD) or a PIN diode. In either device the modulated light signal is converted to an electrical signal which is then amplified and, if necessary, decoded. In decoding, specific time slices or slots are checked for the presence (a binary 1) or absence (binary 0) of a light pulse. These binary digits are combined to form a digital word. In a reverse system to the sampling circuit, digital words are then used to reconstruct the analog wave.

In the system shown in Fig. 1.1, the distance between light source and light detector is short enough that the signal requires no intermediate amplification or regeneration.

When the link becomes too long, the fiber will attenuate the light waves traveling down it so that the light waves cannot be distinguished from noise. In the early days of fiber optics, this attenuation would occur after only a few meters. Recent advances have extended the range to several kilometers before amplification is necessary.

Even with the highest-intensity light sources and the lowest-loss fibers, the light waves finally become so weak or dim from absorption and scattering that they must be regenerated. At this point a *repeater* must be placed in the circuit. This device consists of a light receiver, pulse amplifier and regenerator, and a light source. Together they rebuild the pulses to their former level and send them on their way.

Now that we have seen how fiber optics works, let us look at some of its fantastic advantages as compared to coax or twisted-pair communication systems.

1.3 ADVANTAGES

In a number of situations fiber optics' advantages are so compelling that they cannot be ignored:

1. Extremely wide bandwidth
2. Smaller-diameter, lighter-weight cables

3. Lack of crosstalk between parallel fibers
4. Immunity to inductive interference
5. Potential of delivering signals at a lower cost

But these are merely the *primary* advantages; there are also important *secondary* advantages:

1. Greater security
2. Greater safety
3. Longer life span
4. High tolerance to temperature extremes as well as to liquids and corrosive gases
5. Greater reliability and ease of maintenance
6. No externally radiated signals
7. Ease of expansion of system capability
8. Use of *common* natural resources

Let us examine first the primary advantages and relate them to the world of telecommunications.

1. An *extremely wide bandwidth* means that a greater volume of information or messages or conversations can be carried over a particular circuit. Whether the information is voice, data, or video, or a combination of these, it can be transmitted easily over optical fibers. Fiber-optic systems have been placed on the market with bandwidths up to 3300 MHz [3]; the ultimate bandwidth may be as high as 10,000 MHz [4].

With such bandwidths it is possible to transmit thousands of voice conversations or dozens of video signals over the same circuit. But more than clearing congestion on the nation's communications channels, it will allow communications services we could not afford until now. Just as one example, direct memory-to-memory transfer between computers will now be possible even as computer speeds double [5].

2. *Smaller-diameter, lighter-weight cables* are obvious advantages with the hair-thin optical fibers. Even when such fibers are covered with protective coatings, they still are much smaller and lighter than equivalent copper cables. For example, a 0.005-in. diameter optical fiber in a jacket about 0.25 in. in diameter can replace a 3-in. bundle of 900 pairs of copper wire [6]. As shown in Fig. 1.3, this enormous size reduction (easily 10:1) allows fiber-optic cables to be threaded into crowded underground conduits or ducts. In some cities these conduits are so crammed they can scarcely accommodate a single additional copper cable.

Figure 1.3   Size reduction with optical fibers. (Courtesy of Corning.)

The size reduction makes fiber-optic cables the ideal transmission system for ships, aircraft, and high-rise buildings, where bulky copper cables take up too much space.

Together with the reduction in size goes an enormous reduction in weight: 208 lb of copper wire can be replaced by 8 lb of optical fiber [7]. It is an important advantage in aircraft, missiles, and satellites.

The combined advantages of smaller-diameter cable and lighter weight give a decided cost savings in transportation and storage. For instance, the Army has found that a 1¼-ton trailer is sufficient to transport optical fibers of the same capacity as three 2½-ton truckloads of copper cable [6].

Because it is so thin and lightweight, fiber-optic cable can be handled and installed much easier (and with less cost) than copper cable. Thus, there is no need to dig up city streets to lay new conduits.

3. In conventional communication circuits, signals often stray from one circuit to another, resulting in other calls being heard in the background. This *crosstalk is negligible* with fiber optics even when numerous fibers are cabled together.

4. *Immunity to inductive interference.* As dielectrics, rather than metal, optical fibers do not act as antennas to pick up radio-frequency interference (RFI), electromagnetic interference (EMI), or electromagnetic pulses (EMP). The result is noise-free transmission. That is, fiber-optic cables are immune to interference caused by lightning, nearby electric motors, relays, and dozens of other electrical noise generators which in-

duce problems on copper cables unless shielded and filtered. Carrying light rather than electrical signals, fiber-optic cables ignore these electrical disturbances. Thus, they can operate readily in a noisy electrical environment. They are particularly useful in nuclear environments because of their immunity to EMP effects. Because of their immunity to electromagnetic fields, fiber-optic cables do not require bulky metal shielding and can be run in the same cable trays as power cabling if necessary.

5. *Potential of delivering signals at a lower cost.* Sand, the basic ingredient of glass optical fibers, and plastic are of course cheaper than copper. However, because of the ultrapurity needed and the relatively low volume of production at present, individual fibers cost $0.30 to $1.50 a yard, compared with about $0.02 a yard for a pair of copper wires [8]. As fiber-optic systems become more common, the price is expected to drop significantly. But for now, designers must closely examine the cost-to-benefit ratio.

In very short applications, it is difficult for fiber optics to compete economically with copper wires. However, where the communication capacity would require coax rather than copper wires, or where interference would require special shielding for metallic wire, fiber links can be competitive even at today's prices [9].

Lifetime costs for a fiber-optic system may be much more attractive and a better basis for comparing fiber optics with wire pairs or coax. Such costs include shipping, handling, and installation as well as manufacture. Before the cable is installed, shipping and handling costs are about one-fourth that of current metal cable and labor for installation is about one-half less [10].

According to a report by International Resource Development Inc., as costs for petroleum and petroleum-related products rise, the use of fiber optics becomes even a more favorable alternative to copper cable since less plastic coating is needed for fiber-optic cable. Still another cost saver, IRD points out, is that connectors used with copper cable are usually gold-plated, whereas fiber optic connectors are made from nylon and plastic [11].

Because of low line losses, fewer line amplifiers (repeaters) are needed for fiber optics. This will reduce system cost as well as maintenance cost.

Harris Corporation reports that its project for Mountain Bell Telephone will be 30% under the cost of a comparable conventional system [12].

Aside from twisted wire pairs and coaxial cable, fiber optics' only other competitor is microwave transmission. Although microwaves can reliably transmit as many bits per second of data, fiber optics promises to be far less costly than microwave towers [2].

By themselves these *primary* advantages are sufficient to justify the use of fiber optics in a number of applications. However, the *secondary* advantages must not be overlooked.

1. *Greater security through almost total immunity to wiretapping* is a matter of much greater importance to military services, banks, and computer networks than it is to the average citizen who is calling a relative thousands of miles away. But for these groups, communication security, that is, telephone or data privacy, is well worth any increased cost.

   Unless a steel cable is added to a fiber-optic cable for strength, a fiber cable can be laid in an undetectable fashion. It just cannot be found with metal detectors or electromagnetic flux measurement equipment as is the case with wire pairs and coax.

   As the light in an optical fiber does not radiate outside the cable, the only way to eavesdrop is to couple a tap directly into the fiber. If an eavesdropper was smart enough to do this, he or she could force some light (and therefore the message) out, but the loss would be so great at the receiving end that an alarm would be sounded. When that happens, one could measure the distance to the tap within a few inches by using time-domain reflectometers, which are discussed in Chapter 8.

   At least that is the theory, although some experts disagree slightly. Elion and Elion, for example, note that it is technically feasible to tap signals from the fiber cladding surface to obtain legible signal-to-noise levels. However, by the use of various modulation schemes on irregular schedules with multiplexing and scrambled signal transmission, the best fiber tapping scheme would be made useless [10].

2. *Greater safety* is available with fiber optics because only light, not electricity, is being conducted. Thus, if a fiber-optic cable is damaged, there is no spark from a short circuit. Consequently, fiber-optic cable can be routed through areas (such as chemical plants and coal mines) with highly volatile gases without fear of causing fire or explosion. In effect, as long as the fiber-optic cable does not have a steel strength member it provides electrical isolation between the transmitter and the receiver. If a cable is disrupted, there can be no short circuits or circuit-loading reflections back to the terminal equipment [2]. In addition, there is no shock hazard with fiber-optic cables. Fibers can be repaired in the field even when the equipment is turned on.

3. A *longer life span* is predicted for fiber optics: 20 to 30 years, compared to 12 to 15 years for conventional cable [12]. Glass, after all, does not corrode as metal does.

4. Because fiber-optic cables are made of glass or plastic, in contrast to metal, they have a *high tolerance to temperature extremes as well as to liquids and corrosive gases*. The cables used at Lake Placid for the Winter Olympics were tested to temperatures down to $-55°$ C.

5. *Greater reliablity and ease of maintenance* is made possible by the extended distance (or spacing) between line amplifiers (repeaters) which boost signal strength. Transmission losses are lower in fiber-optic cable than in coaxial cable, allowing substantial increases between repeaters. Instead of placing a repeater each mile, as in conventional copper wire and coaxial cable systems, repeaters can be positioned 4 miles or more apart. (The objective in present research is to place them 10 miles apart.) With such distances it is often possible, at least in metropolitan areas, to transmit between telephone exchanges without the use of a single repeater. Obviously, the fewer repeaters there are in a circuit, the less likelihood there will be for circuit failure. The reliability increases accordingly.

   In addition to reducing the number of repeaters, the few that are left can be placed indoors. Where the active elements are located within the switching offices, Bell Canada engineers point out, the ease in locating trouble and access to failed equipment will be extremely beneficial to improving services and lowering cost.

6. As fiber-optic cables do not radiate signals, fiber-optic transmission does not interfere with other services. *Signal confinement* is excellent.

7. Many fiber-optic systems can be easily updated to *expand system capability*. By simply changing LED light sources to injection lasers as they become available, one can upgrade most present-day low-loss fiber-optic systems without replacing the original cable. Improved modulation techniques can accomplish the same goal.

8. By using a *common* natural resource—sand—rather than a scarce resource—copper—fiber optics is helping to conserve a dwindling world resource.

1.4 APPLICATIONS

With all these advantages, there are obviously a lot of places, a wide range of fields, where fiber optics can be used to advantage. But rather than consider possibilities, let us examine some systems that are presently in operation in the field, not just in the laboratory.

By the time this is published, numerous additional systems will have been placed into operation, so that any statistics we give now will be out of date. Nevertheless, it is important to note that at present 100 major fiber-optic systems are in operation around the world, according to Information Gatekeepers, Inc. Most of these systems are for public use, with military, government agencies, and industry making up the balance of the users.

Fiber-optic systems are now being used for telecommunications, computers, cable television, military electronics, and industrial instrumentation.

1.4.1 Telephone Systems

The world's first optical link providing regular telephone service to the public was placed in operation on April 22, 1977, by General Telephone Company of California. Since installation it has provided high-quality voice circuits without interruption. The system connects the company's long-distance switching center at Long Beach with its local exchange building 5.6 miles (9 km) away in Artesia. Two repeaters approximately 2 miles apart are used. Light sources are light-emitting diodes; light receivers are avalanche photodiodes. Pulse code modulation is used. The cable contains six fibers: two for voice circuits, two for spares, and two for testing. This allows 24 simultaneous telephone conversations (a bit rate of 1.544 Mb/s), which is far less than the system's capacity.

In Nevada, a fiber-optic cable connects two Central Telephone Company offices in Las Vegas, which are 2.5 miles (4.2 km) apart. The system uses a laser diode but no repeaters. The cable has six fibers which can carry up to 1800 telephone conversations.

In Tampa, Florida, eight optical fibers connect General Telephone of Florida's main switching center with a central office 3.9 miles (6.3 km) away. Only two of the fibers are now in operation, carrying up to 672 telephone calls. As communications traffic increases, four more fibers will be activated for telephone calls and two will be put into use to carry local television signals. Just as in the Nevada system, no repeaters are needed.

Probably the longest telephone link in operation is a 32-mile (50-km) fiber-optic cable between Calgary and Cheadle in southwestern Alberta, Canada. Built by the Harris Corp. for Alberta Government Telephone, the system operates at the rate of 4032 phone conversations for each of its five operating fibers. At this rate, the system will transmit 274 million bits of digital information per second (Mb/s). At its ultimate capacity, the system will be able to handle 20,160 telephone calls simultaneously. It can also be used to transmit television programs and computer data. The light source is an injection laser diode; the light receiver is an avalanche photodiode. Repeaters are spaced at 2-mile intervals.

These are just some of the telephone systems now using fiber optics. Numerous others are now under construction.

1.4.2 Cable Television

Fiber optic links are being used to transmit video signals (1) between TV cameras and vans, (2) in closed-circuit TV surveillance, (3) as parts of regular network TV links, and (4) in cable TV systems.

Since 1976, 34,000 cable TV subscribers in Hastings, England, have received TV signals over a 4760-ft (1.4-km) fiber-optic link installed by Rediffusion Ltd.

In the United States, Suffolk Cablevision (Central Islip, New York) has a 3-km fiber-optic link for transmitting video signals between its receiving site and its studio facilities. In Joplin, Missouri, Cablecom has a 3.56-mile fiber-optic cable connecting their local origination studio with their head-end equipment. It is now carrying four TV channels but is capable of as many as 60. It is said to be the longest continuous aerial fiber-optic TV cable in the world.

In London, Ontario (Canada), a 7.8-km fiber-optic trunk connects a cable TV head-end to a hub distribution center. With a bandwidth of 322 Mb/s, the trunk is designed to carry 12 video channels and 12 FM stereo channels. The ½ in. cable trunk has eight fibers. It is said to be the world's first combination of digital television transmission and optical fiber transmission to go into regular cable service [13].

In the small town of Higashi-Ikoma, near Osaka, Japan, the Japanese government is providing financial backing for a two-way (interactive) cable TV system using fiber optics. The system is called Hi-Ovis (Higashi-Ikoma Optical Visual Information System).

In addition to conventional TV programs, Hi-Ovis provides a variety of services, from local marriage announcements to a library of video programs. Only 158 homes are now connected to the system, but many millions of dollars are being spent on the project.

Each home has a keyboard and controller, microphone, black-and-white television camera, and a television receiver. This equipment is connected via two optical fibers to a central facility which originates local programs, provides stored programs on request, and rebroadcasts nine regular TV channels. One fiber carries one TV channel to the home while the other fiber transmits video signals back to the central facility [14,15].

1.4.3 Power Stations

Fiber-optic systems are now being used to provide telephone and data communication into and within power stations at the Minnesota Power and Light Company, Florida Power and Light Company, and Georgia Power Company. When copper cables are used for such communication, ground potential rise, induction, and high-frequency arc noise are encountered. But with fiber optics these problems are avoided, as the optical links are simply insensitive to interference [16].

1.4.4 Computers

In Houston, Texas, a 4900-ft fiber-optic cable is being used to transmit data to five video terminals in the city's main library from a Sperry Univac computer in the Municipal Courts building. This bidirectional system replaced conventional telephone links using modems [17].

Probably the oldest computer system application for fiber optics is in

Bornemouth, England, where a 10-Mb/s link connects command-room video terminals and a main computer used by Dorset County police. It was installed in 1975 after lightning disabled the display system [18].

In New York City, New York Telephone Co. is testing a fiber-optic system that connects computers in two of the company's buildings. Whereas computers usually communicate at the rate of 50,000 b/s, this system allows transmission speeds up to 44 Mb/s. Eventually, the system is expected to have a speed of 274 Mb/s. The accuracy of the system is expected to be 10 billion times better than the current acceptable error rate of 1 error per 10,000 bits on wire cable [19].

1.4.5 Military

In one of the first military uses of fiber optics, the U.S. Navy installed a fiber-optic system on the cruiser U.S.S. *Little Rock* for shipboard voice transmission. It was successfully deployed in the Mediterranean for more than 3 years [20].

In the Airborne Light Optical Fiber Technology (ALOFT) program, the Navy demonstrated the use of fiber optics in flight tests on board an A-7 test aircraft [21]. Using fiber-optic cables, total cable and connector weight was reduced from 31.9 lb to 2.7 lb [20].

In 1977, the Air Force successfully demonstrated a fiber-optic link between two tactical command and control communications centers. The dramatic weight and volume reduction made possible by fiber optics in this link showed how military transport requirements could be relieved [7].

1.4.6 Miscellaneous Uses

The following systems are just part of the uses of fiber optics for communications:

1. Noncontact temperature measurements
2. Monitoring of current and voltage at high-power stations
3. Barbed-wire perimeter fence that warns of intruders
4. Industrial process control of pit furnaces
5. Measuring hazardous high-level electromagnetic fields without perturbing the field (and thereby distorting the measurements)

1.5 WORLDWIDE USE

Fiber-optic systems are now being used or installed in Argentina, Australia, Belgium, Canada, China, Denmark, France, Great Britain, Israel, Italy,

Japan, Netherlands, Singapore, the United States, the USSR, and West Germany. International interest is keen.

Major expenditures are being made in the United States, Japan, Canada, and Great Britain. In research and development in fiber optics, Japan is the big spender. In a recent year, probably 60% of research and development funds abroad were spent in Japan, exclusively on civilian programs. In fact, Japan's expenditures for civilian R&D in fiber optics surpass the funds spent on civilian R&D in the United States [22].

With the assistance of the Japanese government, Japanese industry is developing significant, exciting new devices and transmission equipment for fiber optics. Repeater spacings as long as 62 km have been demonstrated in the field and as long as 100 km in the laboratory. With such spacing, significant cost savings are possible [23].

In Saskatchewan, Canada, a 3200-km fiber-optic system is being built to carry cable TV signals to the province's 51 largest communities, about 190,000 homes. Later the system will carry voice and data signals. Each fiber in the 12-fiber cable will carry one video channel or 672 one-way voice circuits. In the first stage of installation, Regina and Yorkton will be connected by a 200-km underground link [24].

In Britain, the British Post Office (BPO) is collaborating with British industry to build an optical-fiber network which is predicted to be the most comprehensive of its kind in the world. Nearly 280 miles (450 km) of cable will be installed on 15 routes in England, Wales, and Scotland. It is part of a major drive by the BPO to speed the adoption of optical-fiber communications in Britain [25].

1.6 HISTORY

Although the transmission of information by light waves over glass or plastic fibers is a relatively new invention, communication by light through the atmosphere is indeed a very old process. In fact, at the end of the sixth century B.C., the news of Troy's downfall was passed by fire signals via a chain of relay stations from Asia Minor to Argos [26].

Centuries later in the United States, American Indians were using smoke signals for communication, and Paul Revere was watching for lantern signals. But by the 1790s, progress was being made in optical communication. Claude Chappe built an optical telegraph system on hilltops throughout France. By means of semaphores, messages reputedly could be transmitted 200 km in just 15 minutes [27]. For its time, it was quite an invention, but by the middle of the nineteenth century, Chappe's telegraph had been replaced by Morse's electric telegraph.

In 1880, Alexander Graham Bell carried his invention of the telephone one step further: instead of transmitting sound waves over wires he used a beam of light. Although Bell considered his photophone to be one of his better inventions, the contraption of mirrors and selenium detectors was cumbersome and unreliable. Also, it had such a limited range that it had no practical value.

Much later, in 1927, Baird in England and Hansell in the United States proposed the use of uncoated fibers to transmit images for television, but their ideas were not pursued [28, p. 1].

It was in the 1930s that single-glass-fiber optics were first used for image transmission [29], but the phenomenon was more a laboratory stunt than a practical system.

In the 1950s, studies by A. C. S. van Heel of Holland and H. H. Hopkins and N. S. Kapany of England led to the development of the flexible fiberscope, which is widely used in medical fields [28, p. 1]. Kapany coined the term *fiber optics* in 1956 [28, p. 2].

Optical fibers were developed and placed on the market in the 1950s as light guides which enabled people to peer into otherwise inaccessible places, whether the interior of the human body or the interior of a jet engine. The amount of light lost in these fibers was fantastic, but for the few feet involved in most applications, it did not matter. Still there was no serious consideration given to telecommunication via these fibers.

With the invention of the laser in 1960, research into optical communication was greatly accelerated. The laser's high carrier frequency promised a tremendously wide bandwidth for transmitting information. In theory, a single laser beam could carry several thousand television channels or many many thousands of telephone conversations. Excitement was high. But even though the laser's output could be focused into an extremely narrow and intense beam, it was found that fog and rain could interrupt this beam as it was sent through the atmosphere. Because of this interference, it was actually easier to transmit a reliable laser signal from Arizona to the moon than between downtown and uptown Manhattan [27]. Line-of-sight transmission through the atmosphere, it was concluded, was impractical.

In an attempt to develop a transmission line for laser beams, Charles K. Kao in 1968 proposed low-loss optical fibers for such lines. At that time, fiber losses were very high—more than 1000 dB/km. Then in 1970, Kapron, Keck, and Maurer of Corning Glass Works announced the development of optical fibers with losses of less than 20 dB/km [30]. It was a very significant breakthrough.

Suddenly, long-distance telecommunications by fiber optics was possible. More refinements were made to the fibers; cable losses were then cut from 20 dB/km to 6 dB/km and then lower. Room-temperature semiconductor lasers, LEDs, connectors, and photodiodes were developed for use with these low-loss optical fibers. Cable prices began to drop. Within 6 years, working systems were demonstrated and placed on the market.

Rediffusion of London installed probably the first commercial system in 1976 to transmit TV signals to its cable TV subscribers. That year Bell Labs began a major laboratory demonstration of fiber optics for telephones at its Atlanta, Georgia, facility. The system worked over a distance of 10.9 km without the use of repeaters. Each fiber in the 144-fiber cable could handle 672 telephone calls [31].

The first commercial telephone use of fiber optics was the following year in Long Beach, California, as we discussed previously.

Since that time developments have been so numerous that a person would have to read dozens of journals and reports each month to keep up with the history of this technology.

1.7 LOOKING AHEAD

Fiber optics is indeed a major new invention. Its future is exciting, particularly if it overcomes some of its present limitations. A lack of standardization of fiber-optic components makes interchangeability difficult. Multiple input/multiple output systems are not yet practical. And costs are still significant, although going down almost every day. For these reasons, it is safe to say that fiber optics will not replace all coaxial cables and twisted wire pairs this year or any time soon.

With fiber optics, it should be noted, a wired nation is technically possible. Picturephones, an unlimited number of TV programs, electronic games, and two-way cable TV are just a few of the services that will be possible. Some predict a futuristic world where we will be flooded with information arriving on a fiber-optic cable.

But even though such services may be technically possible, there still remain nontechnical problems such as cost, convenience, and government regulations. For example, it may be technically possible to receive and print a newspaper in your home via fiber optics. But it would be very inconvenient to store blank paper for such a machine and it would be a nuisance to load the machine every day.

Broadband telecommunications being tested in Canada and Japan may never be tried in the United States because of government regulations. In these systems, it makes sense to transmit telephone, data, and television signals into a home or office over a *single* fiber-optic cable. For this to happen in the United States at least, telephone companies would have to go into the cable TV business, or vice versa. Under present U.S. laws, this would be extremely unlikely.

Fiber optics may make numerous services available at a reasonable cost, but it has not yet been established that the public will be willing to pay even modest charges for some of the far-out services being promoted.

Just as with many other inventions, uses may be found for fiber optics that no one can now foresee. Whatever their nature, if they are to be widely adopted they must not only offer services people can afford but also services they want.

To predict the size of these markets is just as risky as trying to determine their uses. Nevertheless Fig. 1.4 is presented to show a market prediction by system application. By the year 2000, as much as 10 trillion kilometers of fiber could be installed worldwide, predicts Koji Kobayashi, chairman of Japan's Nippon Electric Company [32].

The price of multifiber cable is expected to drop from 1978's price of $1 per conductor meter to $0.10 per conductor meter for high-volume orders in the early 1980s [33].

In the more immediate future, let us look at some major systems being planned for automobiles, military, telephone systems, and nuclear power stations.

Fiber-optic systems have long been used in automobiles to guide light to panel displays and to indicate lamp failure. A more important use for fiber optics in automobiles will be to simplify the heavy wiring harness used to transmit signals to accessories. In this use, fiber optics would cut costs, avoid electromagnetic interference, and reduce weight. General Motors may start such production by 1983 [15].

For the military, a 15,000-km fiber-optic communication network is being built for the intercontinental ballistic missile, MX. This highly secure net will include 5000 communications and data-processing nodes and thousands of optical repeaters. It is being built by GTE Sylvania, Inc., for the Air Force [34].

Bell Laboratories is planning an underwater fiber-optic cable to stretch 6500 km between the United States and Great Britain and the Continent. The system is expected to cost only one-third as much as today's undersea cable system. It will carry 4032 conversations per fiber, compared with 200 for a copper wire. Repeaters will be spaced 35 km apart, rather than 9 km, the present standard. Each repeater will have an operating laser and three standby lasers to keep the system's mean time before failure at 8 years [35].

One of the most significant applications of fiber optics is a 611-mile link planned by the American Telephone and Telegraph Company to connect Washington, Philadelphia, New York, and Boston by 1984. AT&T and eight Bell System operating companies on this route have filed applications with the Federal Communications Commission to build this $79 million system. When completed it will carry up to 80,000 simultaneous telephone calls over an 0.5-in. diameter fiber-optic cable. The system may be expanded to Chicago and to Miami later in this decade [36].

At Los Alamos Scientific Laboratories in New Mexico, fiber-optic cables are being installed in the Antares High-Energy Gas-Laser Facility. Antares is a fusion system which is expected to make more energy than it consumes. The data links for the facility will be in areas of high electromagnetic

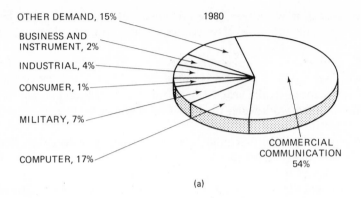

OTHER DEMAND, 15%

BUSINESS AND
INSTRUMENT, 2%

INDUSTRIAL, 4%

CONSUMER, 1%

MILITARY, 7%

COMPUTER, 17%

1980

COMMERCIAL
COMMUNICATION
54%

(a)

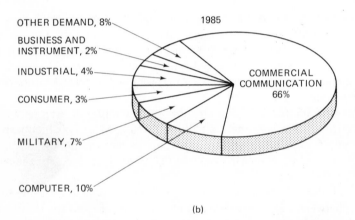

OTHER DEMAND, 8%

BUSINESS AND
INSTRUMENT, 2%

INDUSTRIAL, 4%

CONSUMER, 3%

MILITARY, 7%

COMPUTER, 10%

1985

COMMERCIAL
COMMUNICATION
66%

(b)

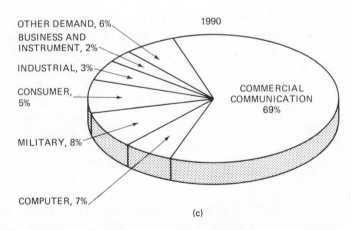

OTHER DEMAND, 6%

BUSINESS AND
INSTRUMENT, 2%

INDUSTRIAL, 3%

CONSUMER,
5%

MILITARY, 8%

COMPUTER, 7%

1990

COMMERCIAL
COMMUNICATION
69%

(c)

Figure 1.4 Fiber-optic systems applications. (From Ref. 2, p. 25; copyright 1978 Mechanical Engineering.)

interference and in addition must withstand high radiation and a pressure/vacuum environment. Copper wires could not be used in such areas without expensive shielding and filtering [37].

1.8 OPPORTUNITIES FOR TECHNICIANS

To be a part of this new technology, to operate, test and maintain these fiber-optic systems, you need to learn the basics of fiber optic components and systems.

The remainder of this book has been prepared to give you the necessary foundation for this challenging innovation. Starting with a discussion of the fundamentals of light, we proceed in a logical fashion from transmitter to receiver to describe components and systems.

REFERENCES

1. "Communications—Bits and Birds Catch On," *Electronics,* Jan. 3, 1980, pp. 129, 136.
2. Fritz Hirschfeld, "The Expanding Industry Horizons for Fiber Optics," *Mechanical Engineering,* Jan. 1978, pp. 20–26.
3. Bob Margolin, "Fiber Optic Links Come of Age," *Electronic Products Magazine,* Nov. 1979, pp. 40–47.
4. R. Andrews, A. Fenner Milton, and Thomas G. Giallorenzi, "Military Applications of Fiber Optics and Integrated Circuits," *IEEE Transactions on Microwave Theory and Techniques,* Dec. 1973, p. 764.
5. James A. Bollengier, "State of the Art of Fiber Optics and Its Applications to the Shipboard Data Multiplexer System," Thesis, Naval Postgraduate School, Monterey, Calif., Mar. 1979, p. 86.
6. "Introduction to Fiber Optics and AMP Fiber-Optic Products," HB 5444, AMP Incorporated, Harrisburg, Pa., n.d.
7. C. W. Kleekamp and B. D. Metcalf, "Fiber Optics for Tactical Communications," Mitre Corp., Bedford, Mass., Apr. 1979, p. 13.
8. Gene Bylinsky, "Fiber Optics Finally Sees the Light of Day," *Fortune,* Mar. 24, 1980, p. 116.
9. Ira Jacobs, "Lightwave Communications—Yesterday, Today, and Tomorrow," *Bell Laboratories RECORD,* Jan. 1980, p. 7.
10. Glen R. Elion and Herbert A. Elion, *Fiber Optics in Communications Systems* (New York: Marcel Dekker, Inc, 1978).
11. "Gold, Copper Costs to Spur Fiber Optic Connector Market?" *Electro-Optical Systems Design,* Apr. 1980, p. 16.
12. News Release, Harris Corporation, Melbourne, Fla., Jan. 16, 1980 (?).

13. "Two Leaps Ahead for Cable: London, Ontario Trunk Puts Together Fiber Optics and Digital Video," *Broadcast Management/Engineering,* Oct. 1978.

14. *Microwaves,* Jan. 1980, p. 58.

15. "GM Plans 400-Hz Fiber Link for Cars to Cut Costs, Weight and Interference, *Laser Focus,* Jan. 1980, p. 60.

16. C. A. Ebhardt, "FTS-1, T1 Rate Fiber Optic Transmission System for Electric Power Station Telecommunications," paper presented at the Second International Fiber Optics and Communications Exposition, Chicago, Sept. 5-7, 1979.

17. "Optical Fiber Cable," Belden Corp. Application Report No. 1-6/9, n.d., Geneva, Ill.

18. Ira Jacobs and Stewart E. Miller, "Optical Transmission of Voice and Data," *IEEE Spectrum,* Feb. 1977, pp. 33-41.

19. "New York Tel Begins Test of Lightwave Data Link," *Telephony,* May 12, 1980, p. 11.

20. "Fibre Optics—Expanding Fleet Capability," Naval Ocean Systems Center Technical Document 224, Mar. 1979.

21. John D. Anderson and Edward J. Miskovic, "Fiber Optic Data Bus," Northrop Corp., Hawthorne, CA and ITT Cannon, Santa Ana, CA, n.d.

22. "Fiberoptics Review and Outlook 1980, R&D Dominates Fiber Spending Abroad; Japan Surpasses U.S. Civilian Efforts," *Laser Focus,* Jan. 1980, p. 65.

23. Sadakuni Shimada, "Optical Systems: A Review—II. Japan: Unusual Applications," *IEEE Spectrum,* Oct. 1979, pp. 70-77, specifically pp. 74-76.

24. "3,200-km Fiber Systems in Saskatchewan Will Be Backbone of Integrated Network," *Laser Focus,* May 1980, p. 64.

25. R. J. Raggett, "BPO Begins Installation of Fiber Optic Network," *Telephony,* May 26, 1980, p. 49.

26. P. Russer, "Introduction to Optical Communications," in *Optical Fibre Communications,* ed. M. J. Howes and D. V. Morgan (Chichester, England: John Wiley & Sons Ltd., 1980), p. 1.

27. W. S. Boyle, "Light-Wave Communications," *Scientific American,* Aug. 1977, p. 40.

28. N. S. Kapany, *Fiber Optics, Principles and Applications* (New York: Academic Press, Inc., 1967), p. 1.

29. *The Optical Industry and Systems Directory, Encyclopedia Dictionary,* 24th ed., The Optical Publishing Co., Inc., Pittsfield, Mass, 1978.

30. Steward E. Miller, Enrique A. J. Marcatili, and Tingye Li, "Research toward Optical Fiber Transmission Systems," *Proceedings of the IEEE,* Vol. 61, No. 12, Dec. 1973, pp. 1703-1751; from p. 1704.

31. Kenneth J. Fenton, "Fiber Optics: Claims Are Being Substantiated," *Information Display,* Jan. 1979.

32. "NEC Chairman Predicts Installation of 10^{10} km of Fiber by Year 2000," *Laser Focus,* Jan. 1980, p. 56.

33. "New Firm Looks to $100 Million Market," *Electronics,* Feb. 16, 1978, p. 50.

34. James Brinton, "Sylvania to Build 15,000 km Fiber-Optic Net," *Electronics,* Feb. 14, 1980, p. 47.

35. Harvey J. Hindin, "Fiber-Optic Cable to Go to Sea for Phone Company," *Electronics,* Mar. 13, 1980, pp. 39–40.

36. Victor Block, "Bell System Files with FCC for D.C. to Boston Fiber Link," *Telephony,* Feb. 4, 1980, p. 11.

37. *DuPont Industry News,* Feb. 27, 1979.

2

FUNDAMENTALS
OF LIGHT

As many electronics technicians have had no education or experience in optics, in this chapter we discuss those fundamental characteristics and phenomena of light that relate to fiber optics. We limit the discussion to those principles of light that are relevant to a basic understanding of optical fibers, light-emitting diodes, injection lasers, and photodetectors. These principles are necessary for understanding fiber-optic components discussed later. Detailed mathematical explanations of lenses, quantum mechanics, and the like will not be given, as they are needed more by the physicist than by the technician.

2.1 NATURE OF LIGHT

2.1.1 Wave vs. Particle Theory

For centuries, physicists have attempted to describe the nature of light, to provide us with a simple picture or model, just as the model of the atom helps us to understand electronics. But there is no simple description, no single theory that explains it completely. However contradictory it may sound, light has the characteristics of particles and of waves.

At times, light seems to be a stream or rain of fast-moving electromagnetic particles called *photons*. Although they are called particles, they bear little resemblance to material particles. For example, they move at fantastic speeds, but at rest they have zero mass. That is, a photon at rest is nonexistent. Instead of calling them particles, perhaps it would be better to call them discrete bundles or packets of energy.

Using the *particle theory,* physicists can describe what happens to light when it is emitted or absorbed. In particular, this theory explains the photoelectric effect, when light striking the surface of certain solids causes them to emit electrons. Without this theory, the behavior of light during emission and absorption cannot be adequately described. But it still does not explain many other phenomena of light.

For example, in numerous experiments light appears to behave like electromagnetic waves instead of a stream of photons. These electromagnetic waves consist of oscillating electric and magnetic fields. Each field is at right angles to each other and to the direction of travel or propagation.

The strength of each field varies sinusoidally. Because these fields oscillate at right angles to the direction of propagation, light waves are said to be *transverse.* Like other electromagnetic waves, light can travel through empty space and over very great distances.

The *wave theory* best explains light propagation or transmission. It also explains why light beams can pass through one another without disturbing the other. Notice what happens when two searchlight beams cross each other. Each proceeds from the intersection as if the other had not been there [1]. With particles, this phenomenon could not occur.

Consider, too, the phenomenon of interference. If light from a single source is split into two beams and if the two beams travel over two unequal paths to reach a common point, the beams will interfere with each other. Depending on their phases, these beams will either increase or decrease the intensity of the light at the common point. This characteristic can only be explained by the wave theory.

Therefore, to describe the nature of light completely we must use both the particle theory and the wave theory. Most often we will use the wave theory, but remember that neither theory is sufficient in itself. Both theories can be used, depending on the problem. But to make a perfect mental picture or model of light is just not possible at this time.

2.1.2 The Wavelength of Light

If light is an electromagnetic wave, what then is its frequency or wavelength?

In most electronics work it is customary to refer to electromagnetic waves in terms of *frequency* rather than *wavelength.* In optics, however, wavelength is the important term, as it can be measured directly.

The wavelength of light waves is most typically measured in nanometers (nm), micrometers (μm), and angstroms (Å), instead of feet or inches. (In some literature, meter may also be spelled "metre.") The relationship of these units is as follows:

Unit	Old Designation	Dimensions
micrometer[a]	micron	10^{-6} m or 10^{-4} cm
nanometer[a]	millimicron	10^{-9} m $\quad$ 10^{-7} cm
angstrom		10^{-10} m $\quad$ 10^{-8} cm

[a] 1000 nm = 1 μm.

The old designations, unfortunately, are still being used in some books and articles, even though they are obsolete. All three units are commonly used by optics workers, but *nanometers* is the unit used most commonly in fiber optics. A typical wavelength encountered in fiber optics is 820 nm; this, of course, may be expressed also as 0.820 μm or 8200 Å.

To give you an idea of the extremely small wavelengths involved in fiber optics, consider the following scale:

|++++++++++++++++++++++| 20 mm

The distance between each mark is 1 millimeter. A wavelength of 820 nm therefore would be 820 millionths of this distance, or 0.0000323 inch! Figure 2.1 shows the scale of light-wave communications.

2.1.3 Electromagnetic Spectrum

Fundamentally, there is no difference between light waves and other electromagnetic waves, such as radio and radar, except that light waves are much shorter and therefore have a much higher frequency. When all types of electromagnetic radiation are arranged in order of wavelength, the result is called the *electromagnetic spectrum,* shown in Fig. 2.2.

Notice how broad this range is: from long electrical oscillations with wavelengths measuring thousands of kilometers to cosmic rays with wavelengths in trillionths of a meter. There are no gaps in the spectrum, but some of the regions overlap or blend. That is, the boundary between regions is not sharp.

As can be seen in Fig. 2.2, optical radiation lies between microwaves and x-rays. It includes all wavelengths between 10 nm and 1 mm. Within this range, as shown on the left side of Fig. 2.2, are ultraviolet, visible light, and infrared radiation. The term *visible light* seems redundant; however, it is necessary because ultraviolet and infrared radiation are referred to as ultraviolet light and infrared light, respectively, in some textbooks. *Visible light* is defined as that radiation which stimulates the sense of sight (that is, affects our optic nerves). It includes all radiation from 390 to 770 nm, from violet to red. Light itself does not have color, but these wavelengths stimulate color receptors in the eye. Obviously, the visible spectrum is just a small fraction of the electromagnetic spectrum.

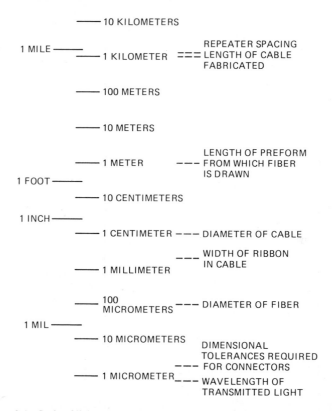

Figure 2.1   Scale of light-wave communications. (From Ref. 2; copyright 1976 Bell Laboratories RECORD.)

In fiber optics, a typical wavelength is 820 nm. From Fig. 2.2 we see that this radiation is designated infrared, although it is sometimes referred to as light because it can be controlled and measured with instruments similar to those used for visible light.

2.1.4 The Speed of Light

Like other electromagnetic waves, light travels through free space (that is, a vacuum) at the fantastic speed of 186,000 miles per second or 300,000,000 (3 × 10⁸) meters per second. For precision work, the figure of 2.997925 × 10⁸ meters per second (m/s) should be substituted for this figure. But for most practical or ordinary uses, the round number of 3 × 10⁸ m/s is satisfactory. Light traveling in the atmosphere moves slightly slower than this, but the figure of 3 × 10⁸ m/s is still sufficiently accurate.

For propagation in free space and in the atmosphere, the speed of light is the same for all wavelengths; however, in other materials, such as water and

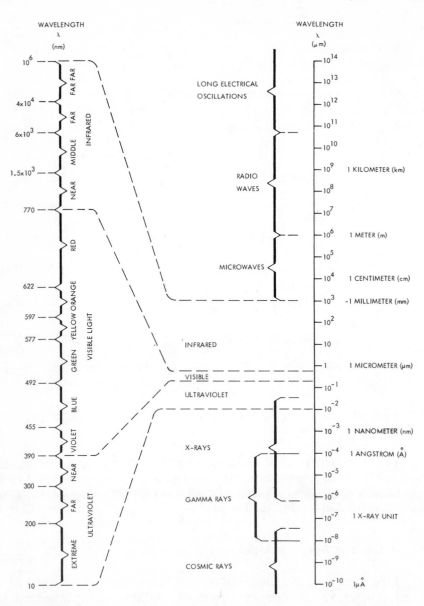

Figure 2.2 Electromagnetic spectrum. (From Ref. 3; copyright 1974 RCA Corporation.)

glass, different wavelengths travel at different speeds. Regardless of the wavelength, when light travels through such materials, its speed is noticeably reduced. Because of this slowing down, a light ray moving from air into a solid or a liquid will bend at the surface of the new medium. (In optics, a medium is any substance that transmits light.) Such bending or refraction is very important to the study of fiber optics, so it will be discussed in more detail in a later paragraph.

The speed of light has an importance far beyond the field of optics: It is one of the fundamental constants of nature. As such, it crops up in numerous scientific equations, some of which have nothing to do with light. In equations it is designated as the constant c.

2.1.5 Straight-Line Propagation

For most practical purposes, such as carpentry and navigation, light can be considered to travel in an essentially straight line as long as it stays in a uniform medium. (A uniform medium is simply a substance that has uniform composition throughout.)

Any bending of a light ray traveling through the atmosphere is so slight that it can be ignored in most measurements that are based on light. Thus, light is said to have the property of *rectilinear propagation* when it is traveling through a uniform medium.

When it travels from one medium to another, however, it will change its direction or bend at the boundary between the two mediums. Just how much it will bend is discussed later in the section "Refraction." But once it enters a second uniform medium, it will continue in a straight line as long as it stays in that medium.

2.1.6 Transparent, Translucent, and Opaque Materials

Liquids, gases, and solids can be classified as transparent, translucent, or opaque, depending on how much light can penetrate or pass through them. If light can pass through a material with little or no noticeable effect, the material is called *transparent*. In this category we would place water, air, some plastics, and glass. The glass used for optical fibers is so ultrapure, according to Bell Laboratory scientists, that if seawater were so clear, you could see to the bottom of the deepest ocean.

If no light can pass through a material, the material is said to be *opaque*. But even transparent materials can become opaque if you increase the thickness or number of layers sufficiently. Notice that clear water at the edge of a lake where the water is shallow is transparent for a few feet from the shore. In the center of the lake, however, at depths of hundreds of feet, no light can pass through.

Some items—wax paper, for example—will pass light but you cannot see clearly through them. These items are called *translucent*.

2.1.7 Rays of Light

In designing and analyzing optical equipment, it is often desirable to draw simple diagrams showing lenses and other optical devices. But how should the path of the light through the equipment be drawn?

Consider a small light source that is radiating light waves in all directions. These waves can be pictured as spherical surfaces concentric with the source. As the distance of the wave from the source increases, the curve begins to flatten out, forming a straight line. Thus, when the distance becomes large enough, the spherical surfaces can be considered planes. The result is a train of plane waves [4].

A train of wavefronts can be drawn as shown in Fig. 2.3. The drawing

INCIDENT
WAVES

MEDIUM 1
MEDIUM 2

Figure 2.3 Train of wavefronts.

can be further simplified by replacing the train of light waves with a single straight line called a *ray* (Fig. 2.4), which is drawn in the direction in which the waves are traveling.

Optical problems can often be solved using such rays and determining their angles and relationships through the use of geometry. This approach is called *geometrical optics*. As it is the simplest approach, it is widely used in the design of optical equipment.

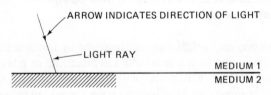

ARROW INDICATES DIRECTION OF LIGHT

LIGHT RAY
MEDIUM 1
MEDIUM 2

Figure 2.4 Light ray.

In contrast, *wave optics* considers only the wave properties of light in the treatment of optical problems.

2.2 REFLECTION

When a light beam strikes an object, some or all of the light will bounce off or be turned back; the light is said to be *reflected*. If the surface of the object is smooth and polished, as with a sheet of silver, *regular* reflection or *specular* reflection will occur, as shown in Fig. 2.5.

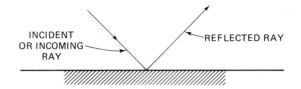

Figure 2.5   Regular or specular reflection.

If the surface of the object has irregularities or is rough in comparison to the wavelength of light, as is usually the case, the light will be reflected in many directions, as illustrated in Fig. 2.6. This case represents *diffuse* reflection.

In everyday use, the ordinary mirror illustrates specular reflection, whereas most nonluminous bodies demonstrate diffuse reflection. Without diffuse reflection we would be unable to see such bodies.

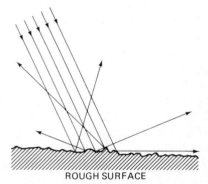

Figure 2.6   Diffuse reflection. (Courtesy of George Shortley and Dudley Williams, *Elements of Physics,* 3rd ed., p. 512; copyright 1961; reprinted by permission of Prentice-Hall, Inc., Englewood Cliffs, N.J.)

Let us now add an imaginary line called the *normal* to Fig. 2.5 and obtain Fig. 2.7. Note that the normal is perpendicular to the surface. The angle *i,* formed by the normal and the incident ray, is called the angle of incidence. The angle *r* is called the angle of reflection. By the *law of reflection,* the angle

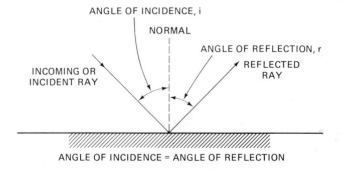

Figure 2.7   Law of reflection.

of incidence equals the angle of reflection. That is, $i = r$. Furthermore, the incident ray, the reflected ray, and the normal are always in the same plane.

2.3 REFRACTION

When light strikes a surface head on, as shown in Fig. 2.8, part of it will be reflected (not shown) and part will be absorbed, as shown by the penetrating ray. We assume in this case that the material will transmit light. For this discussion, we ignore the reflected ray and concentrate on the penetrating ray. As long as the incident ray is perpendicular to the surface, it will continue in a straight line in the new medium, as shown in the figure.

Even though the penetrating ray will not change directions in this particular case, it will slow up appreciably. In free space or air, as you will recall, the velocity of light is 186,000 miles per second. However, in most substances it is less. For instance, in water it is 140,000 miles per second and in an optical fiber it is 124,000 miles per second.

Now consider the case when the light ray is not head on, but at an oblique angle, as illustrated in Fig. 2.9. Instead of incident ray AB continuing in a straight line as BC, it changes its direction to BD. (If the angle of incidence is greater than the critical angle, which will be described later, refraction will not occur.) This bending or refraction is caused by the change of speed of the ray as it enters medium 2. In this particular case, medium 2 is more dense than medium 1 and therefore the refracted ray bends toward the normal. (If medium 1 had been more dense than medium 2, the refracted ray would bend *away* from the normal and line BD would lie on the other side of BC.)

The angle i, formed by the incident ray and the normal, is the angle of incidence. For refraction to take place, this angle must be greater than 0 degrees and less than 90 degrees. The angle θ, formed by the normal and the refracted ray, is the angle of refraction.

Before we look at the relationship of the angle of incidence to the angle of refraction, we need to look at the *index of refraction* (also called the refractive index), which is designated by the letter n. This index is simply the ratio of the speed of light in air (c) to the speed of the light being considered (v), or

$$n = c/v$$

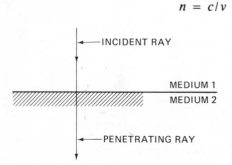

Figure 2.8 Light ray penetrating a second medium.

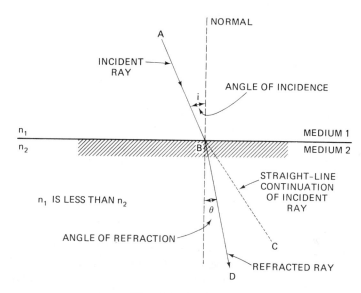

Figure 2.9   Refraction.

Typical indexes of refraction are given in Table 2.1. As we have noted earlier, different wavelengths of light travel at different speeds. The typical indexes are based on a wavelength of 5890 Å, the wavelength of a sodium flame.

When a light ray traveling in a medium with an index of refraction of n_1 strikes a second medium (with an index of refraction of n_2) at an angle of incidence i, the angle of refraction θ can be determined very easily by Snell's law:

$$n_1 \sin i = n_2 \sin \theta \qquad (2.1)$$

Notice that if medium 1 is air, $n_1 = 1$ and can be dropped from the equation, leaving

$$\sin i = n_2 \sin \theta$$

Just as in the case with reflected rays, the incident ray, the normal to the surface at the point of incidence and the refracted ray will all be in the same plane.

TABLE 2.1   Typical Indexes of Refraction

Air	1.00
Diamond	2.42
Ethyl alcohol	1.36
Fused quartz	1.46
Glass	1.5–1.9
Optical fiber	1.5
Water	1.33

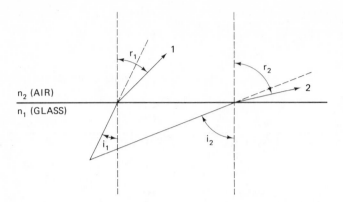

Figure 2.10 Light ray bending away from normal.

In our diagrams up to this point, we have shown the incident ray and the refracted ray going down on the page. However, this path can be exactly reversed and the ray can go up instead of down. In such a case equation (2.1) still applies as long as the incident ray is defined as the ray in the less dense medium and the refracted ray as the one in the more dense medium.

When a ray of light is moving from a dense medium (high index of refraction) to a less dense medium (lower index of refraction) it will *not* be refracted if it strikes the surface at an angle equal to or greater than a particular angle called the *critical angle*. Instead, it will be totally reflected at the surface between the two media.

To see how this phenomenon of total internal reflection occurs, recall that a light ray will bend *away* from the normal if it is moving into a less dense medium, as shown by ray 1 in Fig. 2.10. As the angle of incidence is increased, from i_1 to i_2, notice that the angle of refraction increases so that ray 2 is closer to the boundary between the two media. (Although not shown in Fig. 2.10, reflection also occurs until the critical angle is reached.)

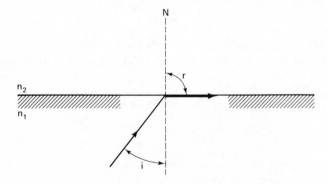

Figure 2.11 Total internal reflection: $i = \theta_c = $ critical angle.

When the angle of incidence is increased to a high enough value, called the critical angle (θ_C), the refracted ray just grazes the surface and travels parallel to it, as shown in Fig. 2.11. At this point, the angle of refraction r is 90 degrees.

Substituting in Snell's law, we obtain

$$n_1 \sin r = n_2 \sin 90 \deg$$
$$= n_2 \, (1)$$
$$\sin r = \frac{n_2}{n^1}$$

But

$$\sin r = \sin \theta_C$$

Therefore,

$$\sin \theta_C = \frac{n_2}{n_1}$$

Thus, for any ray whose angle of incidence is greater than this critical angle, total internal reflection will occur at the surface between two media provided that the light ray is traveling from a medium of higher refractive index to a medium with a lower refractive index.

Dispersion.   Up until this point, we have assumed, although we may not have stated it, that our light beam or ray consisted of only one wavelength. Such light, which is called *monochromatic,* is not naturally encountered in the real world. Most light beams are complex waves which contain a mixture of wavelengths and thus are called *polychromatic.* As shown in Fig. 2.12, white light can be separated into individual wavelengths by a glass prism through the process of *dispersion.*

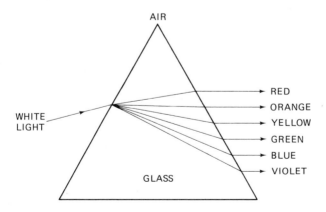

Figure 2.12   Dispersion.

Notice that white light is actually a combination of six colors. Dispersion is based on the fact that different wavelengths of light travel at different speeds in the same medium. Remember that in Table 2.1 we specified a particular wavelength. Because different wavelengths have different indexes of refraction, some will be refracted more than others.

2.4 LIGHT MEASUREMENTS

Two systems of optical measurements are encountered in fiber optics: radiometry and photometry. *Radiometry* applies to the measurement of radiant energy of the entire spectrum, regardless of wavelength. *Photometry* is concerned only with measuring that part of the spectrum that affects the eye. That is, photometry does not consider radiation that the eye cannot see.

Although these systems use different terms and symbols, they are related and conversion can be made from one system to the other, as shown in Tables 2.2 and 2.3.

Numerous terms are *not* included in these tables because there has been little standardization between the various technical societies, industry, and publishers. The distinction between some of the terms is often quite subtle and confusing. And some of the so-called obsolete terms refuse to go away. But by

TABLE 2.2   Radiometric and Photometric Units

| Definition | Radiometric | | Photometric | |
	Name	Unit (SI[a])	Name	Unit (SI[a])
Energy	Radiant energy	joule	Luminous energy	lumen-second
Energy per unit time = power = flux	Radiant flux	watt	Luminous flux	lumen
Power input per unit area	Irradiance	W/m^2	Illuminance	lm/m^2 lux
Power per unit area	Radiant exitance	W/m^2	Luminous exitance	lm/m^2
Power per unit solid angle	Radiant intensity	W/steradian	Luminous intensity	candela
Power per unit solid angle per unit projected	Radiance	$W/m^2 \cdot steradian$	Luminance	$candela/m^2$

[a] International System of metric units—recommended standard.

Source: Ref. 5; courtesy of M. Grossman.

TABLE 2.3 Conversion Tables for Illumination and Luminance

Illumination Conversion Factors[a]

1 lumen - 1/680 lightwatt (at 555 nm)	1 watt-second = 1 joule = 10 ergs
1 lumen-hour = 60 lumen-minutes	1 phot = 1 lumen/cm^2
1 footcandle = 1 lumen/ft^2	1 lux = 1 lumen/m^2

Number of ⟶ Footcandles Multiplied by ⟍ Equals number of ↓	Footcandles	Lux	Phots	Milliphots
Footcandles	1	0.0929	929	0.929
Lux	10.76	1	10,000	10
Phots	0.00108	0.0001	1	0.001
Milliphots	1.076	0.1	1,000	1

Luminance Conversion Factors

1 nit = 1 candela/m^2
1 stilb = 1 candela/cm^2
1 apostilb (international) = 0.1 millilambert = 1 blondel
1 lambert = 1,000 millilamberts

Number of ⟶ Foot-lamberts Multiplied by ⟍ Equals number of ↓	Footlamberts	Candelas /m^2	Milli-lamberts	Candelas /in.^{2}	Candelas /ft^2	Stilbs
Footlamberts	1	0.2919	0.929	452	3.142	2.919
Candelas/m^2	3.426	1	3.183	1,550	10.76	10,000
Millilamberts	1.076	0.3142	1	487	3.382	3,142
Candelas/in.^{2}	0.00221	0.000645	0.00205	1	0.00694	6.45
Candelas/ft^2	0.3183	0.0929	0.2957	144	1	929
Stilbs	0.00034	0.0001	0.00032	0.155	0.00108	1

[a] Footcandle is an obsolete term.

Source: Ref. 5; courtesy of M. Grossman.

going back to the basic units—lumens and watts—you can resolve many of the conversion problems.

The *watt* is the fundamental unit in radiometry and the *lumen* is the fundamental unit in photometry. At the eye's peak sensitivity, at a wavelength of 555 nm, 1 watt = 680 lumens.

The 16th General Conference on Weights and Measures has defined the *candela* as the luminous intensity, in a given direction, of a source emitting monochromatic radiation of frequency 540 x 10^{12} Hz and whose radiant intensity in this direction is 1/683 watt per steradian [6].

A steradian is defined in Fig. 2.13. Figure 2.14 shows some of the relationships of the terms in Table 2.2. Figure 2.15 shows light-level brightness values of common objects.

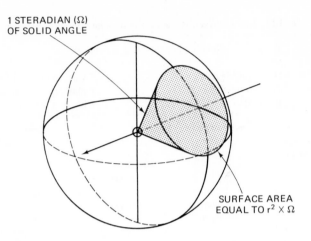

Figure 2.13 Steradian. (From Ref. 5; courtesy of M. Grossman.)

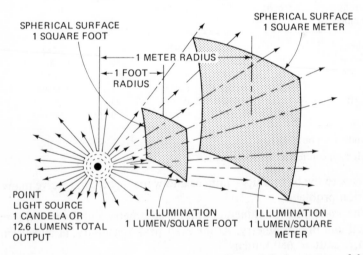

Figure 2.14 Illumination by a point source. (From Ref. 5; courtesy of M. Grossman.)

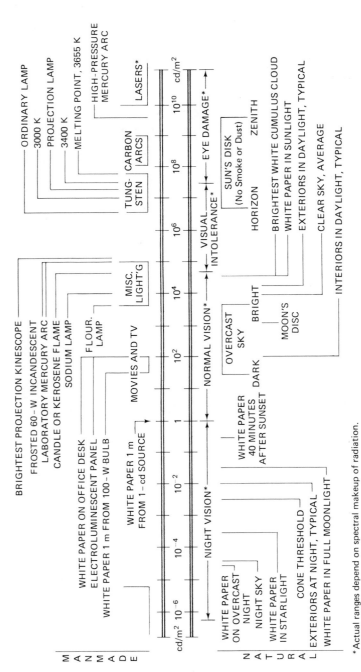

Figure 2.15 Luminance ranges. [Reprinted with permission from Ref. 7 (*Electronic Design*, Vol. 26, No. 4); copyright Hayden Publishing Co., Inc., 1978.]

*Actual ranges depend on spectral makeup of radiation.

REFERENCES

1. Physical Science Study Committee, *Physics* (Lexington, Mass.: D. C. Heath and Company, 1960), p. 257.

2. Ira Jacobs, "Lightwave Communications Passes Its First Test," *Bell Laboratories RECORD*, Dec. 1976, p. 297.

3. *RCA Electro-Optics Handbook*, p. 14.

4. Francis Weston Sears, *Optics* (Reading, Mass.: Addison-Wesley, 1949), p. 4.

5. Morris Grossman, *Technician's Guide to Solid-State Electronics,* (Englewood Cliffs, N.J.: Prentice-Hall, Inc., 1976).

6. "Photometric Units Redefined," *Electro-Optical Systems Design,* Apr. 1980, p. 16.

7. "Focus on Lamps," *Electronic Design,* Vol. 26, No. 4, Feb. 15, 1978, p. 58.

3

LIGHT SOURCES AND TRANSMITTERS

Numerous devices are available for converting electronic signals to light waves for fiber-optic telecommunication systems. However, at present only two of these devices are really suitable for fiber optics: the light-emitting diode (LED) and the injection laser. Both are semiconductor diodes which are directly modulated by varying the input current. They are close relatives, both being made of aluminum–gallium–arsenide (AlGaAs or GaAlAs).

The LED used in fiber optics is similar to those LEDs used in numerous everyday world applications, especially in visual displays such as pocket calculators and some digital wristwatches. It is sometimes referred to as an "infrared-emitting diode (IRED)" because some LEDs emit an invisible beam at an infrared wavelength.

The injection laser is also referred to as an injection laser diode (ILD), laser diode (LD), and diode laser (DL).

At present, injection lasers far outsell LEDs in dollar terms; but since injection lasers are so much more expensive than LEDs, they account for less than 10% of the number of light sources sold [1]. The majority of long-distance fiber-optic links use injection lasers, but the majority of systems in use are short-haul rather than long-haul.

Both devices have unique advantages. The final choice between them in any given application depends on cost, optical power levels, modulation rates, wavelength, temperature sensitivity, coupling efficiency, and lifetime. These factors will be considered briefly before we discuss each device in detail.

LEDs and injection lasers for fiber optics are very small devices, as

ACTUAL SIZE

Figure 3.1 LED and injection laser. (Cross-section views are shown in later illustrations.)

shown in Fig. 3.1. At these small sizes they more nearly match the dimensions of the fibers they are coupled to.

Ideally, for high efficiency all light from the light source should enter the fiber. But as shown in Fig. 3.2, LEDs spew light in all directions. Notice how much more concentrated or directional are the emissions from the injection laser. Because of this narrower emission angle, injection lasers have less coupling losses than LEDs. That is, their coupling efficiency is higher.

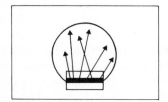 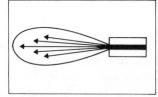

Figure 3.2 Radiating characteristics of an LED (left) and an injection laser (right). (From Ref. 2; courtesy of AEG-Telefunken.)

With present-day injection lasers, tens of milliwatts can be coupled into an optical fiber. In comparison, a typical LED will couple about 10 dB less power into a fiber. Thus, the injection laser couples 10 times as much light into a fiber as does an LED.

In comparing the output of these devices, note that the standard level of reference is 1 milliwatt (mW), which is often expressed as 0 dBm to indicate 0 dB above the standard level. A typical continuous-wave laser has an output of 5 mW but a typical LED's output is 0.5 mW with the same driving power (200 mA at 1.5 V).

Power output of LEDs is most typically measured in *microwatts* (such as 50 to 500 μW); that of injection lasers, in *milliwatts* (such as 1 to 40 mW). As development of these devices continues, these figures are likely to increase. AEG's V213P LED, for instance, is said to have an output of 1 mW [3].

Because of its greater power, the injection laser is best suited for long-distance transmissions, especially when repeaters must be placed far apart.

Along with this major advantage, injection lasers have some noticeable disadvantages:

1. Cost is much higher for lasers than for LEDs: typically hundreds of dollars for lasers, as low as $5 each for some LEDs.

2. Lasers are extremely sensitive to temperature. The output of lasers must be monitored and feedback control must be provided to maintain the output power constant despite temperature variations. These control circuits make the system more complex and therefore less reliable.

3. The lifetime of lasers at room temperature is much less than that of LEDs. Many manufacturers will guarantee a lifetime of only 10,000 hours. However, this figure is likely to change soon as improvements are made. Fujitsu has announced the development of injection lasers with an estimated operating lifetime of 5.7×10^5 (100,000) hours, as a result of accelerated life tests [4]. In contrast, LEDs have operating lives of 10^6 to 10^7 hours [5].

Obviously, none of these devices has been operated for 10^5 hours, as that represents hundreds of years. However, accelerated life tests allow operating life to be predicted. Different accelerated life tests are used by the manufacturers, so that lifetime figures are not always comparable.

A lifetime of 1,000,000 (10^6) hours seems a fantastic time, scarcely necessary. But when several sources—either LEDs or injection lasers—are in a system, such lifetimes are necessary if mean-time-before-failure is to be held to a reasonable figure.

In the LED there is nearly a direct proportion—a linear relationship—between current input and light output. Because of this characteristic, LEDs are ideally suited for transmitting *analog* signals. Especially is this true for short distances and low modulation rates (less than 50 MHz).

At higher modulation rates, however, analog modulation becomes impractical because of pulse dispersion (signal distortion).

Both LEDs and injection lasers may be used to transmit *digital* signals. As the injection laser has a faster response time than the LED, it has a higher modulation rate—typically from 50 megabits per second (Mb/s) and up. (Notice that digital modulation rates are expressed in megabits per second, rather than in megahertz. The relationship between the two rates is discussed in section 3.3.)

Thus, injection lasers are preferred for broad-bandwidth (above 50 Mb/s) fiber-optic links. Modulation in excess of 4 GHz has been achieved [6].

First-generation LEDs and injection lasers transmit at wavelengths between 815 nm and 910 nm, that is, from visible red to invisible near-infrared. Because optical fibers attenuate some wavelengths more severely than others, its is imperative that the wavelength of the source correspond with little deviation to the low-attenuation region of the fiber. Fortunately, present-day fibers exhibit low attenuation in the range 815 to 910 nm. Also, high-efficiency photodetectors are available that are sensitive to these wavelengths.

Second-generation light sources made of indium–gallium–arsenide–

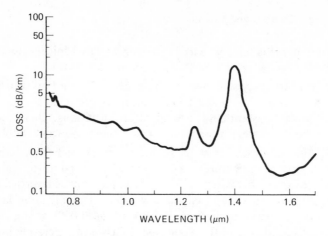

Figure 3.3 Optical transmission losses vs. wavelength. Optical-transmission losses peak at 1400 nm in single-mode fibers used in the far-infrared region. Minimum losses occur at about 1300 and 1500 nm. Some of the most sophisticated systems, as a consequence, are designed for 1550 nm. [Reprinted with permission from Ref. 7, p. 54 (*Electronic Design,* Vol. 28, No. 2); copyright Hayden Publishing Co., Inc., 1980.]

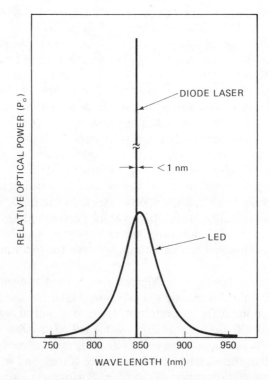

Figure 3.4 Typical emission spectra for injection lasers and LEDs. Typical emission spectra for diode lasers and LEDs show the relatively narrow bandwidth of a DL above the lasing threshold. [Reprinted with permission from Ref. 5 (*Electronic Design,* Vol. 28, No. 8); copyright Hayden Publishing Co., Inc., 1980.]

phosphide (InGaAsP on InP), [7, p. 53] operate in the region 1100 to 1700 nm (far infrared). At these wavelengths there is significantly lower fiber loss, as shown in Fig. 3.3, and less material dispersion (pulse spreading—see Section 4.6).

As shown in Fig. 3.4, the spectrum of injection lasers is much narrower than that of LEDs. With broad-spectrum sources, different wavelengths will have different velocities. These variations in velocity can broaden the pulse excessively, reducing the modulation rate. The narrow spectra of the injection laser means less material dispersion and therefore higher possible modulation rates.

Because of the lower losses at these frequencies, LEDs again become useful, even for long-distance circuits.

3.1 LIGHT–EMITTING DIODES

By now most electronics technicians have become familiar with the small size, low temperature, and ruggedness of conventional LEDs (see Fig. 3.5).

Although similar to these LEDs, fiber-optic LEDs are more complex, precise, and carefully made; and they emit more intense light [8].

As a matter of review, recall that in certain semiconductor diodes, if the *p-n* junction is forward-biased, some of the electrons injected across the junction will recombine with holes. If certain materials and dopants are used, a

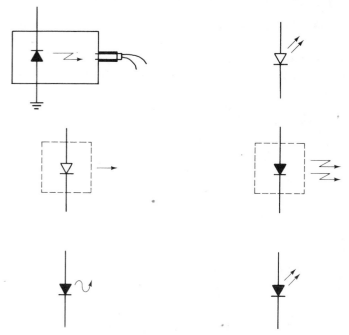

Figure 3.5 Typical symbols for LEDs.

photon will be produced each time an electron falls into a hole, as shown in Fig. 3.6. A ray of light will be formed if a great number of these photons can escape from the diode. The resulting spontaneous radiation is said to be incoherent because the photons are out of step, in random order.

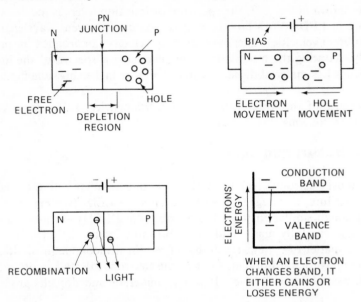

Figure 3.6   Simplified theory of LED operation. (From Ref. 8; courtesy of AMP.)

The materials used in constructing LEDs determine whether the radiation will be visible or invisible (infrared), and if visible, what color. Gallium–aluminum–arsenide (GaAlAs), for instance, will produce invisible (infrared) radiation, while gallium arsenide phosphide (GaAsP) will generate visible (red) radiation.

Which type should be used depends on the optical fiber and the light receiver to be used. Optical fibers have less attenuation for some wavelengths than others, and light receivers or detectors are more sensitive to some wavelengths than to others.

The light power of the LED is approximately proportional to the injection current, but the LED is never 100% efficient. This is because in some electron–hole recombinations there are no photons produced.

LEDs can be modulated by varying the forward current. Some of the best can be modulated up to 200 MHz.

In comparison with lasers, LEDs have good transient and overload protection [7, p. 54].

Two types of LEDs are in common use in fiber optics: surface emitters (developed at Bell Laboratories) and edge emitters (developed at RCA). Surface emitters are more commonly used, primarily because they give better light emission. However, coupling losses are greater with surface emitters and they have lower modulation bandwidths than edge emitters. Both types are discussed in the following paragraphs. Table 3.1 compares some of their characteristics.

3.1.1 Surface Emitters

The most efficient surface emitting LED is the Burrus type, named after its developer. As shown in Fig. 3.7, a well has been etched through the GaAs substrate. This helps to prevent absorption of the radiation. At the same time this well provides a convenient way to bring an optical fiber close to the grain-size light source.

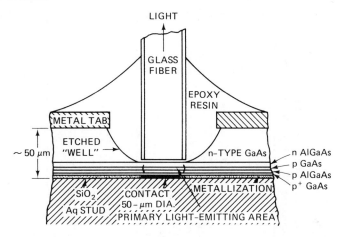

Figure 3.7   Structure of a Burrus surface-emitting LED. (From Ref. 10; copyright 1978 IEEE.)

Because this LED emits light in many directions, it is sometimes necessary to take extra steps to obtain more efficient coupling between the LED and the fiber. This can be accomplished with a microlens (Fig. 3.8) on the surface of the LED chip or a hemispherical dome (Fig. 3.9). These structures are more expensive than the simple surface emitter but allow more power to be coupled into the optical fibers.

Short (12-in.) lengths of fibers called *pigtails* are sometimes bonded to LEDs (and injection lasers) to aid coupling. The pigtail is attached to the light-emitting surface. Once installed, the pigtail allows fiber-to-fiber splicing which can be simpler than aligning the pigtail to the tiny light spot on the LED.

Most often the pigtail is installed by the light-source manufacturer, as an optical bench jig, microscope, and radiometer are necessary for the procedure. It should be noted, however, that some manufacturers feel that non-pigtailed designs are easier to couple.

TABLE 3.1 Characteristics of Surface and Edge-Emitter LEDs: High-Speed Continuous DC (CW) or Pulse-Operated Types

| Type | Chip Material | Wavelength of Peak Radiant Intensity (nm) | Source Size or Fiber-Optic Core Diameter (μm/mils) | Maximum Ratings at $T_C = 27°C$ | | | | |
| | | | | Continuous DC Operation | Pulsed Operation | | | Case |
				DC Forward Current, I_{FM} (mA)	Peak Forward Current, i_{FM} (A)	Pulse Duration, t_w (μs)	Duty Factor du (%)	Operating Temperature Range, T_C (°C)
SG1010	GaAs	940	—	100	10	2	0.2	−40 to +125
SG1010A	GaAs	940	—	100	10	2	0.2	−40 to +125
C30119	GaAlAs	850	25.4 × 62.5/1 × 2.5	200	1.5	0.1	5	−40 to +90
C30122	GaAs	940	—	100	10	2	0.2	−40 to +125
C30123a	GaAlAs	830	25.4 × 62.5/1 × 2.5	200	1.5	0.1	5	−40 to +90
C86011E	InGaAs	1060	62.5/2.5	200	10	0.1	0.1	−40 to +90

TABLE 3.1 (cont.)

Characteristics at $T_C = 27°C$ and the Specified Operating Conditions

Continuous DC Operation					Pulsed Operation			Typical Switching Characteristics
Radiant Flux (Φ) (Power Output)		DC Forward Current I_F	Typical Forward Voltage Drop V_F	Typical Peak Radiant Flux Φ_m (Power Output)	Forward Current, i_F	Pulse Duration, t_w	Pulse Repetition Rate prr	Rise Time
Min.	Typ							
(mW)	(mW)	(mA)	(V)	(mW)	(A)	(μs)	(Hz)	(ns)
2	3.5	100	1.3	26	1	1	500	900
4	7	100	1.3	50	1	1	500	900
0.3	0.5	200	1.5	5	1	0.05	1 MHz	3
0.5	1.0	100	1.3	—	—	—	—	30
0.8	1	200	2	7.5	1	0.05	1 MHz	8
0.15	0.2	200	1.0	—	—	—	—	<10

Source: Ref. 9; courtesy of RCA.
[a] Edge emitter.

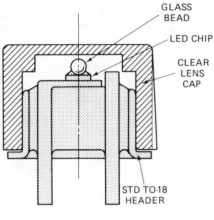

Figure 3.8 Microlens on surface of an LED. A microlens on the surface of the diodes focuses light on the plastic window of Spectronics emitters and detectors. [Reprinted with permission from Ref. 11 (*Electronic Design*, Vol. 28, No. 11); copyright Hayden Publishing Co., Inc., 1980.]

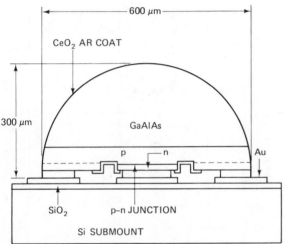

Figure 3.9 Hemispherical dome on LED. The high-power infrared emitting diode from Hitachi uses a forward-biased *p-n* junction to push photons into the GaAlAs crystal dome. The concentration of Al controls the wavelength, and promotes strong emissions. [Reprinted with permission from Ref. 7, p. 59 (*Electronic Design*, Vol. 28, No. 2); copyright Hayden Publishing Co., Inc., 1980.]

The junction of the LED and the pigtail is fragile and may easily break. If this happens, the entire assembly is ruined.

By operating LEDs well below their rated power level—for example, at 20% of maximum drive—their lifetime will go up dramatically, in some cases by a factor of 10 [7, p. 62].

3.1.2 Edge Emitters

As shown in Fig. 3.10, the radiant output of an edge emitter is emitted from the edges of the diode in the recombination region of the junction. An

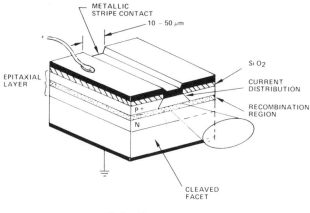

Stripe-Geometry

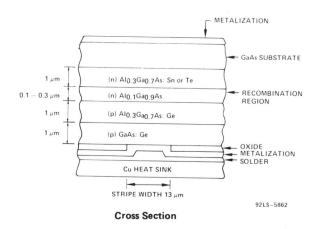

Cross Section

Figure 3.10 Schematic and cross-sectional view of an edge-emitting LED. (From Ref. 9; courtesy of RCA.)

oxide isolated metallization stripe constricts the current flow through the recombination region to the area of the junction directly below the stripe contact. It is possible to confine the radiating portion of the junction to a spot approximately 50 μm in its greatest dimension. Thus, it provides an excellent match to small-diameter fibers [9, p. 3].

3.1.3 Temperature Considerations

Power versus current characteristics vary with ambient temperature for LEDs, but not as drastically as for injection lasers, as shown in Fig. 3.11.

Heat sinks are provided for some LEDs, as shown in Fig. 3.12, to alleviate changes in characteristics because of temperature changes.

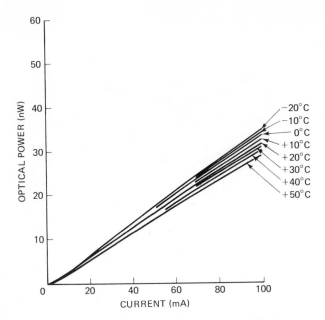

Figure 3.11 Power vs. current characteristics for a GTE laboratory LED. [Reprinted with permission from Ref. 5 (*Electronic Design,* Vol. 28, No. 8); copyright Hayden Publishing Co., Inc., 1980.]

3.1.4 LED Transmitters

Figure 3.13 shows a relatively simple but effective LED transmitter.* The first box in the diagram is the logic interface. For optimum compatibility with data-communication applications, the transmitter must interface with a common logic family and provide some standard load and input-signal requirements. Because by definition it must only handle data rates up to 20M baud, transistor-transistor logic (TTL) is a good choice for the family; standard gates, inverters, and other components can implement the block, providing an electrical input port compatible with any TTL drive level.

The next section of the transmitter's block diagram—the LED driver and current gain—has several functions. Obviously, it must be able to generate enough current to allow the LED to develop its required optical output power. Additionally, it must switch this current on and off in response to the input data, implementing rise and fall times consistent with the maximum expected baud rate. Finally, it must provide enough current gain to amplify the limited source and sink currents available from the logic interface up to the LED's input-level requirements.

In most link designs, the third block in Fig. 3.13—the LED and optical

---

* The following three paragraphs are from Ref. 12; copyright 1980 Cahners Publishing Co., *EDN.*

Figure 3.12   High radiance LED with heat sink. (Courtesy of GTE Lenkurt.)

connector—consists of two separate parts. However, recent product introductions from Motorola and AMP [as well as others] now make it possible to address the electrical-to-optical transducer/fiber-coupling functions simultaneously. Because this single-function concept maximizes coupling efficiency while minimizing mechanical-alignment problems, it is the one employed in the transmitter design.

An experimental LED transmitter designed by Motorola is shown in Fig. 3.14. The transmitter handles NRZ (non return to zero) data rates to 10 Mbits or square-wave frequencies to 5 MHz, and is TTL- or CMOS-compatible, depending on the circuit selected.

Powered from a +5- to +15- V supply for CMOS operation or from a +5-V supply for TTL operation, the transmitter requires only 150 mA of total current. The LED drive current may be adjusted by resistor R1, and should be set for the proper LED power output level needed for system opera-

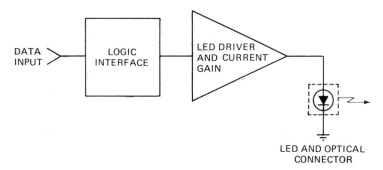

Figure 3.13   Simple LED transmitter (From Ref. 12; copyright 1980 Cahners Publishing Co., *EDN.*)

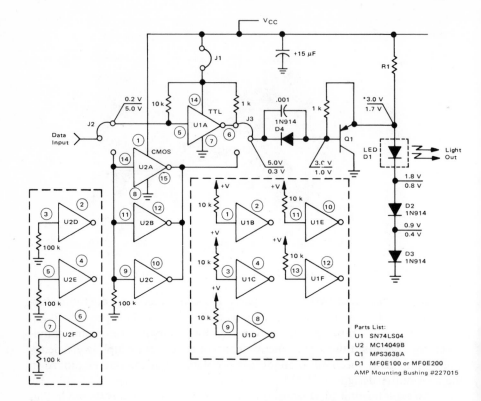

Figure 3.14   Fiber optics LED transmitter connected for TTL operation. (From Ref. 13; courtesy of Motorola, Inc.)

tion. The LED is turned off when transistor Q1 is driven on. Diodes D2 and D3 are used to assure the turn-off. Diode D4 prevents reverse-bias breakdown (base–emitter) of transistor Q1 when the integrated circuits U1 or U2 outputs are high [13].

3.2 INJECTION LASERS

Of the three basic types of lasers—gas, solid, and semiconductor—only the semiconductor laser is practical for fiber optics. This is because of size, voltage, and cost restrictions.

In the semiconductor laser family, the most important device for fiber optics is the injection laser diode (Fig. 3.15), hereafter called "injection laser" or (ILD).

Figure 3.15   Injection laser. (Courtesy of RCA.)

Although more expensive than an LED, the injection laser, as we have noted earlier, can couple higher power into an optical fiber and is ideally suited to high-speed digital systems. Its lifetime is less than that of an LED but is rapidly being improved.

The injection laser is quite similar to the LED. In fact, they are made of the same materials, although arranged somewhat differently. Below a certain threshold current, the ILD acts as an LED—it has spontaneous emission and a broadband light output. Above the threshold current, the laser starts to oscillate, that is, lasing begins, as shown in Fig. 3.16.

When a properly biased current is passed through the ILD (Fig. 3.17), the holes and electrons move into the active region. Some recombine, giving off photons of light in the process. In the LED the photons can escape the die as emitted light, or they can be reabsorbed by the *p* or *n* material. When a photon is reabsorbed, a free electron may be created or heat may be generated. In the ILD something different happens: the light is partially trapped in the active region by the mirrorlike end walls. The photons reflect back and forth. The photon in the active region, as it reflects back and forth, can persuade a free electron to recombine with a hole. The result is a new photon exactly like the first. That is, the first photon has stimulated the emis-

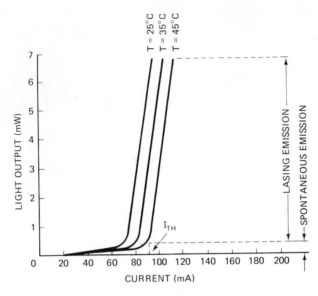

Figure 3.16 Temperature variations vs. light output for injection laser. (From Ref. 14; copyright 1980 Cahners Publishing Co., *EDN.*)

sion of the second. Gain has occurred, for there are two photons where there was but one [8].

For the stimulation to occur, a strong bias current supplying many carriers (holes and free electrons) is required. The current continuously injects carriers into the active region, where the trapped photons stimulate the carriers to recombine and create more photons. The light energy (number of photons) has been pumped up by the carrier injection. This pumping allows amplification [8].

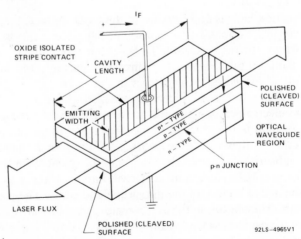

Figure 3.17 Typical injection laser structure. (From Ref. 9; courtesy of RCA.)

All of the light is not completely trapped in the active region; some is emitted from the mirrorlike end surfaces in a strong narrow beam of laser light [8].

The schematic arrangement of typical ILDs is shown in Figs. 3.18 and 3.19. Table 3.2 gives some common characteristics of ILDs.*

RADIANT
FLUX

Figure 3.18   Schematic arrangement of a typical injection laser. (From Ref. 9; courtesy of RCA.)

Figure 3.19   Schematic arrangement of another injection laser. *Note:* Laser chip is at top center. Width of base = 0.220 inches. (Courtesy of General Optronics Corp.)

The forward drive current of an ILD must be held at a constant value above the threshold point to maintain a constant radiant flux output. Threshold current, however, is quite temperature dependent and, as a result, the operating temperature must be stabilized to prevent output drifts.

* The following description of the ILD is from Ref. 9.

TABLE 3.2 Injection Laser Characteristics: Double-Heterojunction Diodes Continuous DC (CW) Operated Types

| Type | Chip Material | Wavelength of Peak Radiant Intensity (nm) | Source Size or Fiber-Optic Core Diameter (μm/mils) | Fiber-Optic Cable | Maximum Ratings at $T_C = 27°C$ | | Maximum Accessible Emission Level[a] (Peak Power Output) (mW) |
					Radiant Flux (Power Output) (mW)	Forward Current (Peak and DC) i_{FM}, I_{FM} (mA)	
C30127	GaAlAs	820	2 × 13/0.08 × 0.5	—	15	400	250
C30130	GaAlAs	820	2 × 13/0.08 × 0.5	—	15	400	250
C86000E[b]	GaAlAs	820	2 × 13/0.08 × 0.5	—	10	200	250
C86014E[b]	GaAlAs	820	2 × 6/0.08 × 0.24	—	10	150	250

With Integral Fiber-Optic Cables and Connectors[b]

Type	Chip Material	Wavelength of Peak Radiant Intensity (nm)	Source Size or Fiber-Optic Core Diameter (μm/mils)	Fiber-Optic Cable	Radiant Flux (Power Output) (mW)	Forward Current (Peak and DC) i_{FM}, I_{FM} (mA)	Maximum Accessible Emission Level[a] (Peak Power Output) (mW)
C86002E	GaAlAs	820	62.5/2.5	Siecor 112	2	300	250
C86006E	GaAlAs	820	100/4	Siecor 1372	3	400	250
C86007E	GaAlAs	820	200/8	DuPont PIFAX-120	6	400	250
C86010E	GaAlAs	820	62.5/2.5	Siecor 112	3	400	250

TABLE 3.2 (*Cont.*)

Characteristics at T_C = 27°C and Specified Operating Conditions

Operating Case Temperature (°C)	Radiant Flux (Power Output) Φ			DC Forward Current to Obtain Specified Minimum Radiant Flux		Typical Threshold Current (mA)	Typical Forward Voltage Drop (V)	Typical Rise Time (ns)
	Minimum (mW)	Typical (mW)	Maximum (mW)	Typical (mA)	Maximum (mA)			
−35 to +50	5	10	15	300	400	250	2	<1
−35 to +50	5	10	15	300	400	250	2	<1
−35 to +50	5	7	10	100	150	75	2	<1
−35 to +50	5	7	10	100	150	75	2	<1
−35 to +50	1[c]	1.5[c]	2[c]	200	400	175	2	<1
−35 to +50	1[c]	2[c]	3[c]	300	400	250	2	<1
−35 to +50	2[c]	4[c]	6[c]	300	400	250	2	<1
−35 to +50	1[c]	2[c]	3[c]	300	400	250	2	<1

Source: Ref. 9; courtesy of RCA.

[a] The maximum peak power output values listed for the different laser types under Maximum Accessible Emission Level are the maximum theoretical levels of radiant flux output obtainable from the devices at a specific wavelength, emission duration, and maximum drive current to which human access is possible. These values are based on product design and include possible changes in device characteristics during life. Appropriate precautions should be taken to avoid harmful exposure.

[b] These devices can be made available with customer supplied fiber.

[c] At output end of integral fiber-optic cable.

The temperature of these devices can be stabilized by using a thermistor sensor in a control loop which includes a small thermoelectric cooler. Such a system provides a constant laser output with a constant drive current over a modest range of temperature. The temperature control should normally be set in the range 18 to 20° C in order to achieve good temperature stabilization.

A typical drive circuit incorporating temperature stabilization is shown in Fig. 3.20. The maximum ratings for radiant flux (power output) or forward current (drive current), whichever is reached first, should never be exceeded.

The user should ensure that the drive circuitry used to power the diode laser is designed to limit the drive current to that required to obtain the maximum specified radiant flux. The laser diode may be operated directly from a power supply. However, before such operation is effected, the power supply should be thoroughly checked for transients.

Exposure of the diode to even very brief transient current spikes can cause irreversible device failure. Safe operating considerations require that the device be protected by connecting a resistor (5 to 10 Ω) in series with the laser diode.

Developments in laser processing and packaging have led to the introduction of laser devices employing integral optical feedback (RCA C86002E).

By employing a silicon PIN photodiode at the rear facet of the laser, a photocurrent proportional to the laser output is generated. This photocurrent may be used in a feedback control loop to provide a stable laser output over a wide range of temperature. Thus, the need for temperature control to obtain a thermally stable laser output is not necessary.

For another typical laser transmitter, let us look at Optical Information Systems OTX5100 Series fiber-optic transmitter, which is a feedback-stabilized ECL-compatible module usable to 275 Mb/s (NRZ). The optical source in this transmitter is an AlGaAs diode laser chip made by Optical Information Systems (OIS). The laser is mounted in the OIS DI-PAC, a metal 14-pin DIP package. The DI-PAC utilizes an Amphenol Type 906 metal connector, which mates with a variety of commercially available optical-fiber cables. The modulated optical power at the end of the connector is greater than 1 mW peak to peak. The laser power is stabilized, using an integral feedback detector mounted at the rear facet of the laser in the DI-PAC.*

Figure 3.21 shows the transfer characteristic of the DI-PAC laser, the electrical input modulation waveform, and the corresponding light-output waveform. The laser is biased to a dc current value (I_{DC}) near the threshold of laser action. The light output at threshold establishes the OFF of baseline power condition. A high-speed pulse driver is used in the transmitter to supply the additional current (I_{MOD}), which increases the light output of the laser to the ON level.

---

* This description of the OTX5100 transmitter is taken from Ref. 15.

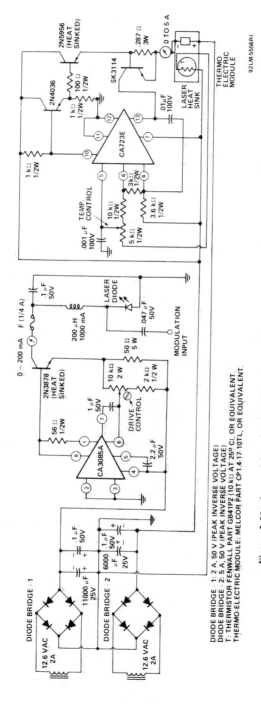

Figure 3.20 Laser drive circuit with temperature stabilization. (From Ref. 9; courtesy of RCA.)

59

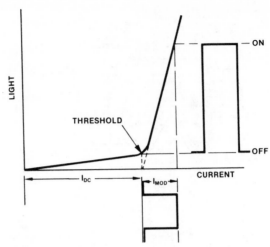

Transfer characteristic of the DI-PAC Laser,
the electrical input modulation waveform, and
the corresponding light-output waveform.

Figure 3.21 Modulated laser characteristics. (From Ref. 15; courtesy of Optical Information Systems, Exxon Enterprises, Inc.)

In the following description of the operation of the transmitter, refer to Fig. 3.22.

The modulation signal input conditioning circuitry provides isolation between the laser and any signal placed at the input terminal. A constant load impedance is presented to the input signal source (50-Ω termination returned to -2 V). The electrical input requirements of the circuitry are the same as those of a standard 10,000 series ECL gate.

The digital voltage waveform from the buffer is used to generate a current-modulation signal for the laser. This high-speed transistor stage supplies the additional current to the laser required to change the light output from the OFF to the ON value (see Fig. 3-21).

The variable bias-current source supplies the dc component of the total laser current. The circuitry contained in this stage is adjusted automatically to maintain the laser baseline power at a fixed value, independent of temperature changes and data format or duty cycle. A set of output terminals are supplied on the transmitter's printed-circuit-board connector to allow the connection of an isolated external voltmeter for bias-current monitoring.

The circuitry in the active current limiter allows the maximum bias current for the laser to be established. As the dc component of the laser current reaches the set value, no further increases in current are possible.

The photodiode current generated by the laser monitor detector mounted in the DI-PAC is amplified and buffered to provide both an external monitor signal (proportional to light output) and a signal to be used for comparison in the error-amplifier portion of the circuit.

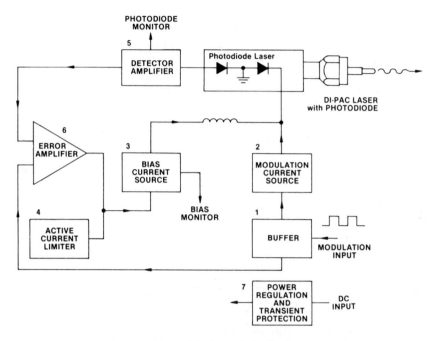

Figure 3.22   Optical transmitter with electronic feedback. (From Ref. 15; courtesy of Optical Information Systems, Exxon Enterprises, Inc.)

Signals from the photodiode detector amplifier and the modulation buffer are combined with the necessary dc components in the error amplifier. The output signal is used to adjust the laser bias current automatically so as to maintain the laser baseline power constant, independent of the modulating signal.

Power supply filtering, decoupling, regulation, and transient protection are included on the transmitter card to protect the laser diode and ensure stable light output.

Injection lasers are being developed which will use indium–gallium arsenide–phosphide (InGaAsP) instead of gallium–aluminum–arsenide and operate at room temperature. The lasers will radiate in the 1200- to 1600-nm band, where optical fibers have their lowest attenuation and distortion. At these wavelengths repeaters can be placed at much greater ranges, tens of kilometers, instead of 7 or 8.

3.3 MODULATION

With both LEDs and injection lasers, *direct* modulation of the light source by varying the current is preferred for most frequencies. *Indirect* or *external* modulation which modifies the light after it leaves the source can be accomplished with electro-optic and magneto-optic modulators. Such modulation is useful for integrated optics (Chapter 9) and it may be the only way to modulate a 7- to 11-GHz link [16]. But for most present uses, only direct modulation is used.

With direct modulation, both analog and digital systems are practical. The increased use of computers in telecommunications is, however, helping the evolution of analog to digital transmission. There is a definite trend to digital because of its flexibility and efficiency.

In digital (pulse) applications, a pulse is formed by turning the source on for a brief instant.* The burst of light is the pulse. Digital implies two states; on/off, 1/0, high/low. These two states represent bits or binary digits. Not only does the burst of light have significance (1), but its absence has significance (0), as shown in Fig. 3.23. A train of pulses or bits is given meaning through any of a variety of coding schemes.

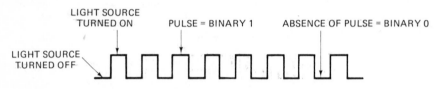

Figure 3.23   Pulse train. (From Ref. 8; courtesy of AMP.)

The following are some of the important terms associated with pulses, as shown in Fig. 3.24:

1. *Amplitude* is the height of the pulse; it is a measure of the pulse's strength.
2. *Width* is the time the pulse remains at its full amplitude.
3. *Rise time* is the time it takes to go from 10% to 90% of the amplitude. In fiber optics, it is related to how long it takes the source to turn on fully.
4. *Fall time* is the time it takes to go from 90% to 10% of the amplitude. It is related to how long it takes the source to turn off fully.

Rise time is perhaps the single most important characteristic in high-speed digital applications, for it determines how many pulses per second are possible. In the two pulse trains in Fig. 3.25, the time slots and the pulse widths are identical. The only thing that allows more pulses in one train is faster rise and fall times.

When we use pulses to represent bits, we speak of speed as a bit rate or so many bits per second. The rising and falling of a pulse is akin to the rising and falling of a sine wave. Bit rate is, in a loose sense, a frequency. Bit rate and sine-wave frequency are related, but they are not the same.

---

* The following discussion of pulses is taken from Ref. 8.

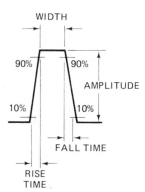

Figure 3.24   Pulse characteristics. (From
Ref. 8; courtesy of AMP.)

Whereas computer engineers specify transmission channel capacity in bits per second, recall that communications engineers express the same capacity in *bauds*. In some code patterns, the terms are interchangeable; in others, the units must be divided, as will be shown, to obtain equivalent units.

In a non-return-to-zero pattern (NRZ), shown in Fig. 3.26, the signal does not periodically return to zero. The signal will remain at the "1" level if the stream of NRZ data contains a series of consecutive "1's." In the same manner, if the stream contains a series of consecutive "0's," the signal will remain at the "0" level. With return-to-zero (RZ) codes, the level periodically changes from high level to low level or back, never remaining at either level for a period of time longer than one bit interval. Since the NRZ code requires only one code interval per bit interval, it uses the channel space most efficiently. Notice that the RZ code uses two code intervals per bit interval [17].

Thus, with NRZ data, 10 megabaud (Mbaud) corresponds to a data rate of 10 Mb/s. With other codes, the data rate is the baud rate divided by the number of code intervals per bit interval [18].

In general, the analog electrical bandwidth required for an NRZ code is one-half the bit rate; for an RZ-coded signal, it is equal to the bit rate. Thus, a 200-MHz bandwidth link can transmit 400 Mb/s of NRZ digital data.

A higher bit rate allows more information to be sent over the line. To show why information-carrying capacity is so important, we look at how large amounts of information are sent over a line. The example is *pulse-code modulation* (PCM), which is used by telephone companies to send several

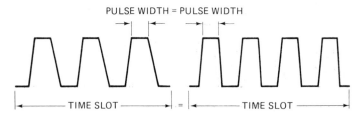

Figure 3.25   Comparison of pulse trains. (From Ref. 8; courtesy of AMP.)

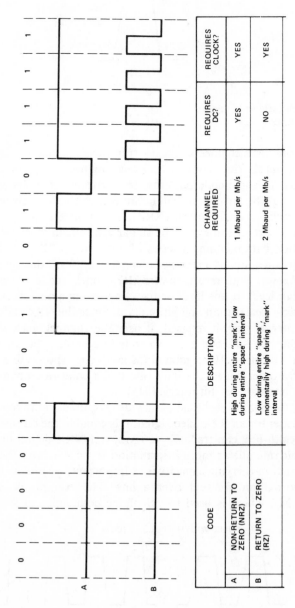

Figure 3.26 NRZ and RZ code patterns. (From Ref. 17; courtesy of Hewlett-Packard.)

	CODE	DESCRIPTION	CHANNEL REQUIRED	REQUIRES DC?	REQUIRES CLOCK?
A	NON-RETURN TO ZERO (NRZ)	High during entire "mark", low during entire "space" interval	1 Mbaud per Mb/s	YES	YES
B	RETURN TO ZERO (RZ)	Low during entire "space", momentarily high during "mark" interval	2 Mbaud per Mb/s	NO	YES

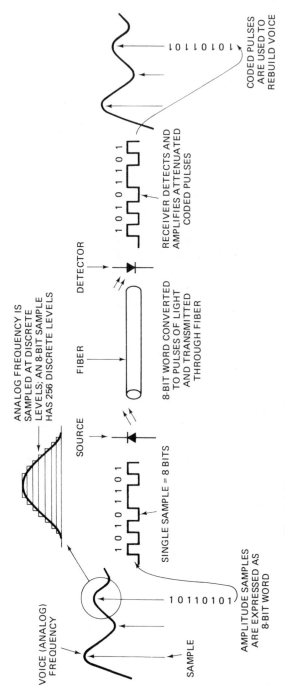

Figure 3.27 Pulse-code modulation (PCM). (From Ref. 8; courtesy of AMP.)

hundred voices over a line. The separation of the voices is done by *multiplexing*. The following discussion is based on a working fiber-optic telephone link.*

In pulse-code modulation, a voice (analog) signal is sampled and converted into a binary code of digital pulses. According to communication theory, a voice can be reconstructed if it is sampled at twice its highest frequency. Since the high end of the speaking voice is 4000 Hz, it must be sampled 8000 times per second. The sampling looks at the amplitude of the voice, and expresses the relative amplitude as a binary number that can be represented as pulses. This number is an 8-bit word. Using 8 bits means that a total of 256 different amplitudes can be coded into a binary form ($2^8 = 256$). So for each voice, we must take 8000 samples every second, and every sample has 8 bits. To transmit the voice, 64,000 bits per second (8000 samples/s x 8 bits/sample = 64,000 bits/s) are required (see Fig. 3.27). If we wish to send two voices over the same line, then double that—128,000 bits/s—are needed .

If we wish to send two voices, we must also have some way to distinguish which pulses belong to which voice. This is done by multiplexing, which may be defined as the simultaneous transmission of more than one channel (or voice) over a line. Actually, the transmission is not totally simultaneous. First one signal is sent, then the second, then the first again, rapidly alternating between signals. In our transmission, we must now add notice of when we switch from one signal to the other. This, too, is done with pulses, and so we need a few more bits than the 128,000 bits/s [8].

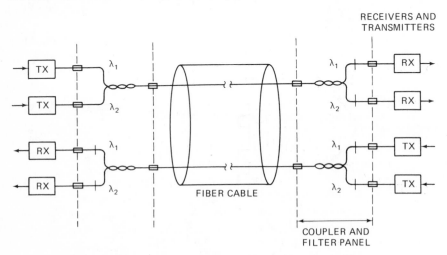

Figure 3.28 Typical approach to wavelength division multiplexing: λ, wavelength; TX, transmitter; RX, receiver. (From Ref. 19; copyright 1979 *Telephone Engineer & Management*.)

* This discussion is taken from Ref. 8.

Three types of multiplexing are common: frequency division multiplexing (FDM), time-division multiplexing (TDM), and wavelength-division multiplexing (WDM). In FDM, signals are interleaved in the radio-frequency domain. In TDM and WDM, signals are interleaved in the optical domain in time and frequency, respectively. Figure 3.28 shows how two or more wavelengths can be operated on a single fiber-optic pair using WDM.

REFERENCES

1. "Sales of Optical Fibers Will Approach 20,000 km as Prices Continue to Drop," *Laser Focus,* Jan. 1980, p. 65.

2. "Fiber-Optic Transmission Technology, Components and Systems," AEG-Telefunken brochure, p. 6, n.d.

3. "Optoelectronic Emitter and Detector Devices to Couple with Fiber Pigtail," AEG-Telefunken Technical Data 1978/1979, p. 4.

4. Fujitsu product literature.

5. R. B. Lauer and J. Schlafer, "LEDs or DLs: Which Light Source Shines Brightest in Fiber-Optic Telecomm Systems?" *Electronic Design,* Vol. 28, No. 8, Apr. 12, 1980, p. 131.

6. T. E. Stockton and R. B. Gill, "Semiconductor Light Sources for Fiber Optic Applications," *Microwave Journal,* Oct. 1980, p. 49.

7. Stephan Orr, "Fiber-Optic Semis Carve Out Wider Infrared Territory," *Electronic Design,* Vol. 28, No. 2, Jan. 18, 1980.

8. "Introduction to Fiber Optics and AMP Fiber-Optic Products", HB 5444, AMP Incorporated, n.d.

9. "Optical Communications Products," RCA Publication OPT-115, June 1979, pp. 12, 13.

10. S. D. Personick, "Fiber Optic Communication: A Technology Coming of Age," *IEEE Communications Society Magazine,* Mar. 1978, p. 15.

11. Gary M. Null, Julius Uradnisheck, and Ronald L. McCartney, "Three Technologies Forge a Better Fiber-Optic Link," *Electronic Design,* Vol. 28, No. 11, May 24, 1980, p. 68.

12. Vincent L. Mirtich, "Designer's Guide to: Fiber-Optic Data Links—Part 1," *EDN,* June 20, 1980, pp. 133–140.

13. "Basic Experimental Fiber Optic Systems," Motorola Advance Information.

14. Joseph F. Svacek, "Transmitter Feedback Techniques Stabilize Laser-Diode Output," *EDN,* Mar. 5, 1980, p. 107.

15. "ECL Compatible Fiber-Optic Transmitter Using a Semiconductor Diode Laser OTX5100 Series," Optical Information Systems, Exxon Enterprises, Inc., Preliminary Data Sheet, Nov. 1979.

16. Douglas Lockie, "I Need More Bandwidth!" *Electro-Optical Systems Design,* May 1980, p. 50.

17. "Digital Data Transmission with the HP Fiber Optic System," Hewlett-Packard Application Note 1000, Nov. 1978.

18. "Fiber Optic 1000 Metre Digital Transmitter," Hewlett-Packard, Jan. 1980.

19. W. R. Fitzgerald and D. F. Hemmings, "Large Role for Fiber Optics Seen for Telcos in Years Ahead," *Telephone Engineer and Management,* Sept. 15, 1979, p. 88.

4

OPTICAL FIBERS

An optical fiber is a thin, flexible thread of transparent plastic or glass which carries visible light or invisible (near-infrared) radiation. Proper choice of this fiber is vital in fiber-optic system design, as the fiber establishes (1) an upper limit to system bandwidth and (2) transmitter-to-receiver or transmitter-to-repeater spacing.

Fiber splicing and connections are discussed in Chapter 5; fiber testing is discussed in Chapter 8. Details of fiber and fiber cable manufacture are not given as technicians are not expected to make their own fiber-optic cables.

4.1 PHYSICAL DESCRIPTION

4.1.1 Fiber Construction

As shown in Fig. 4.1, an optical fiber consists of a central cylinder or *core* surrounded by a layer of material called the *cladding,* which in turn is covered by a *jacket.* The core transmits the light waves; the cladding keeps the light waves within the core and provides some strength to the core. The jacket protects the fiber from moisture and abrasion.

The core as well as the cladding is made of either glass or plastic. With these materials three major types of fiber are made: plastic core with plastic cladding, glass core with plastic cladding, and glass core with glass cladding. In the case of plastics, the core can be polystyrene or polymethyl methacrylate; the cladding is generally silicone or Teflon.

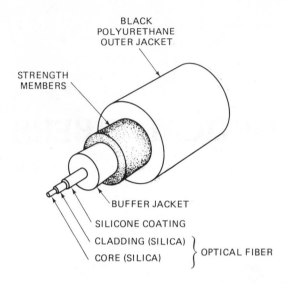

BLACK
POLYURETHANE
OUTER JACKET

STRENGTH
MEMBERS

BUFFER JACKET

SILICONE COATING

CLADDING (SILICA)
                                        } OPTICAL FIBER
CORE (SILICA)

Figure 4.1   Typical optical-fiber construction. (Courtesy of Hewlett-Packard.)

The glass is basically silica, commonly found in sand. Silica occupies 26% of the earth's crust, in stark contrast to copper, which occupies only 0.01% [1]. For optical fibers the silica must be extremely pure; however, very small amounts of dopants such as boron, germanium, or phosphorus may be added to change the refractive index of the fiber. Boron oxide is added to the silica to form borosilicate glass, which is used in some claddings.

To make even more efficient optical fibers, Bell Labs scientists are investigating nonsilicate substances such as zinc cloride. Preliminary results indicate that this material will be 1000 times more efficient than silica fibers [2].

In comparison with glass, plastic fibers are flexible and inexpensive. They are easy to install and connect, can withstand greater stresses than glass fibers, and weigh only 40% as much. However, they do not transmit light as efficiently. Because of their high losses, they are used only for short runs such as within buildings. As glass core fibers are so much more widely used than plastic, subsequent references in this book to fibers will be assumed to be glass, rather than plastic, unless specifically stated otherwise.

In comparison to copper, optical fibers are much lighter: a 40-km fiber core weighs only 1 kg; a 1.4-km copper wire of 0.32 mm outer diameter weighs 1 kg [1].

4.1.2 Dimensions

Optical fibers are typically made in lengths of 1 km (3280 ft) without splices. For special applications, however, some have been made as long as 10,000 ft without splices.

As we will see later, the diameter of the core and the cladding determines many of the optical characteristics of the fiber. The diameter also determines some of the physical characteristics. The fiber must be large enough to allow splicing or attaching of connectors. On the other hand, if it is too large, it will be too stiff to bend, take up too much space in ducts, and use too much material.

An optical fiber is very small, comparable to a human hair. Its outer diameter ranges from 0.1 to 0.15 mm. Compare this with a copper wire, which typically has an outer diameter of 0.32 to 1.2 mm.

Core diameters range from 5 to 600 μm, whereas cladding diameters vary from 125 to 750 μm. To keep the light within the core, the cladding must have a minimum thickness of one or two wavelengths of the light transmitted [3]. The protective jacket may add as much as 100 μm in diameter to the fiber's total diameter. Typical fiber dimensions are given in Fig. 4.2.

At the smallest core diameters (5 μm), severe handling and connection problems are encountered. However, this fiber has a much broader bandwidth.

Within any given fiber, tight tolerances are necessary, as even slight variations in the dimensions can cause significant changes in optical characteristics. Because of this, core diameter tolerances may be specified as low as ± 2 μm.

The International Electrotechnical Commission has proposed two sets of standard fiber dimensions:

1. For graded-index multimode fibers (to be defined later), a core diameter of 50 μm and a cladding diameter of 125 μm.
2. For step-index fibers (to be defined later), a core diameter of 200 μm but an unspecified cladding diameter [5].

All major fiber manufacturers are producing the standard size graded-index fiber, and the Department of Defense has set a similar standard for graded-index fiber [6].

Within any fiber it is desirable to devote as much area as possible to the core rather than to the cladding or jacket. Hence, the highest ratio of core diameter to cladding diameter consistent with core isolation yields the most efficient fiber [7].

4.1.3 Strength

Ordinary glass, as most of us know from experience, is brittle; that is, it is easily broken or cracked. Optical glass fibers, in happy contrast, are surprisingly tough. They have a high tensile strength, that is, ability to withstand hard pulling or stretching. The toughest are as strong as stainless steel wires of

WIDEBAND GRADED INDEX MULTIMODE OPTICAL FIBER
DIMENSIONS SHOWN ARE NOMINAL VALUES

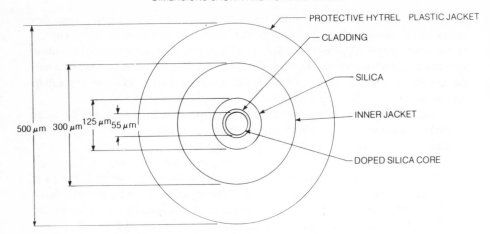

SINGLE MODE STEP INDEX OPTICAL FIBER
DIMENSIONS SHOWN ARE NOMINAL VALUES

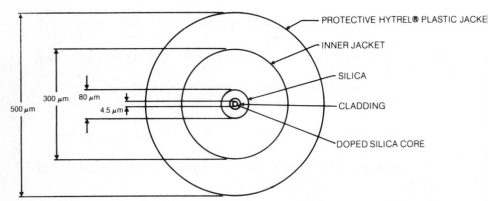

LARGE-CORE PLASTIC-CLAD SILICA OPTICAL FIBER
DIMENSIONS SHOWN ARE NOMINAL VALUES

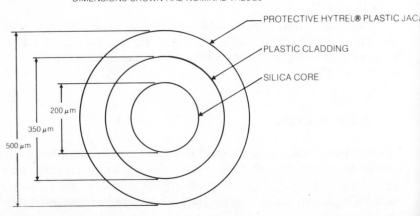

Figure 4.2 Typical optical-fiber dimensions. (From Ref. 4; copyright 1978 International Telephone and Telegraph Corp.)

the same diameter. In comparison with copper wire, optical fiber has the tensile strength of a copper wire twice as thick.

One-kilometer lengths of these fibers have withstood pulling forces of more than 600,000 pounds per square inch before breaking. Bell Labs reports that a fiber 10 m long (about 33 ft) can be stretched by 70 cm (over 2 ft) and still spring back to its original shape [8]. Yet these fibers can be bent into small radiuses (Fig. 4.3)—as low as 2 mm for a 420-μm-diameter fiber. In demonstrations these fibers have been tied in loose knots (not recommended!) without breaking. Only when the knot was drawn tight did the fiber break.

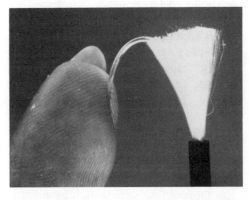

Figure 4.3   Optical fibers can be bent. (Photo courtesy of DuPont.)

To produce fibers this tough, fiber manufacturers try to keep the glass core and cladding free from microscopic cracks on the surface or flaws in the interior (see Fig. 4.4). When a fiber is under stress, it can break at any one of these flaws.

Flaws can develop during and after manufacture. Even a tiny particle of dust or a soft piece of Teflon can give the fiber's surface a fatal scratch [8].

To prevent such abrasion, manufacturers coat the fiber with a protective jacket of plastic (organic polymer) immediately after the fiber is made. This jacket also protects the fiber surface from moisture, which can also weaken

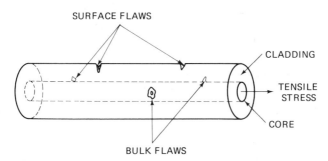

Figure 4.4   Optical-fiber flaws. (From C. K. Kao, "Optical Fibre Cables," in *Optical Fibre Communications*, ed. M. J. Howes and D. V. Morgan; copyright 1980 by John Wiley & Sons Ltd.; reprinted by permission of John Wiley & Sons Ltd.)

the fiber. In addition, the jacket cushions the fiber when it is pressed against irregular surfaces. This cushioning reduces the effect of small random bends (microbends) which otherwise would cause transmission losses. Finally, the jacket compensates for some of the contractions and expansions caused by temperature variations.

Although single-fiber cables are used, generally several fibers are placed together in one cable, as described in Section 4.6. Cables are often designed so that there is little or no stress on the fiber itself.

The maximum tensile strength of a fiber may be specified in giganewtons per square meter, newtons, psi, kg, or MPa. To compare the strength of two fibers, the figures obviously should be in the same units. In addition, the figures should be for the same length of fiber, preferably 1 km.

Splices greatly reduce the mechanical strength of a fiber, as well as causing transmission losses. For these reasons, much effort is made to fabricate continuous fibers with no splices.

4.2 LIGHT PROPAGATION

4.2.1 Refraction and Reflection

Optical fibers guide light by either reflection or refraction, depending on the type of fiber being used. In the reflective types, light rays travel in a zigzag fashion as shown in Fig. 4.5. In the refractive types, light rays travel in a continuous curve (Fig. 4.5). In either case, light rays are confined to the core.

Two reflective types are available: the single-mode step-index fiber and the multimode step-index fiber. Only one type of refractive fiber is available: the multimode graded index fiber.

For the physicist, *mode* is a complex mathematical and physical concept describing the propagation of electromagnetic waves. But for our purposes, mode is simply the various paths light can take in a fiber [10]. By *single mode* we mean that there is only one path for the light; *multimode* means several paths.

To explain *step-index* and *graded-index fibers,* recall that transparent

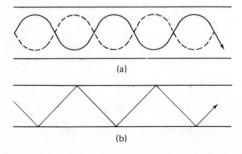

(a)

(b)

Figure 4.5 Methods of optical confinement: (a) refractive confinement; (b) reflective confinement. (From Ref. 9; copyright 1979 T&B/Ansley.)

materials have a refractive index (index of refraction). This optical parameter is designated n and can be computed from the equation $n = c/v$. In this equation, c is the speed of light in a vacuum and v is the speed of light in the material.

From this equation we see that a higher index corresponds to a slower speed of light in the material. Thus, light traveling in a material with $n = 1.48$ will travel slower than in a material with $n = 1.41$.

In optical fibers, the refractive index of the core (n_1) is greater than the index of the cladding (n_2). That is, $n_1 > n_2$.

In a step-index fiber, the refractive index is constant throughout the core. As shown in Fig. 4.6a and b, the index of the core (n_1) is represented by a flat line, parallel to the horizon. Notice in this profile the abrupt change be-

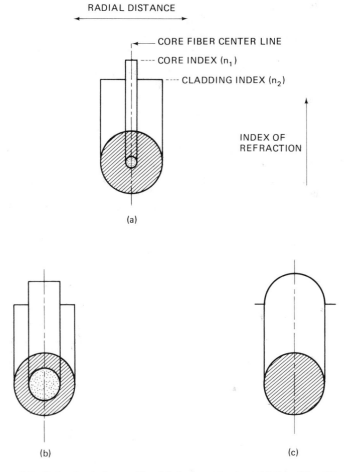

Figure 4.6   Refractive-index profiles: (a) single-mode stepped index; (b) multimode stepped index; (c) graded index.

tween the index of the core and the index of the cladding. This abrupt change gives these fibers the name "step-index."

In a graded-index fiber, the refractive index is not the same throughout the fiber. It is highest at the center of the core but decreases or tapers off radially toward the outer edge, as shown in Fig. 4.6c.

In the single-mode step-index fiber, the core is so small that it allows

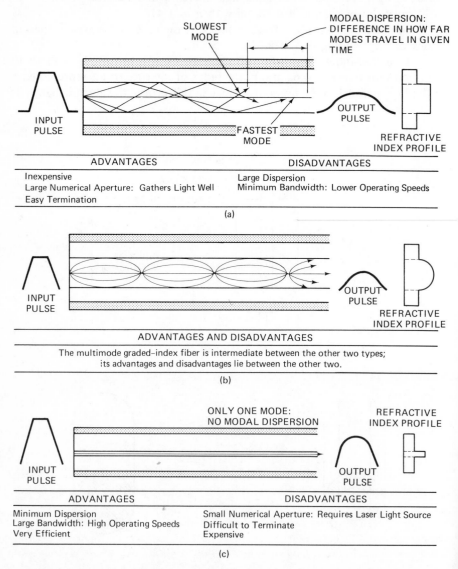

Figure 4.7 Types of light propagation in fibers: (a) multimode step-index fiber; (b) multimode graded-index fiber; (c) single-mode step-index fiber. (From Ref. 10; courtesy of AMP.)

only a single ray to travel down the fiber. This ray in effect travels down the axis of the fiber as shown in Fig. 4.7c.

Within a step-index fiber, light rays from the core strike the cladding at various angles of incidence, as shown in Fig. 4.8. Ray *A*, which is perpendicular to the interface, is transmitted primarily through the interface. Ray *B*, which forms an incident angle ϕ with the normal, also transmits through the interface, but a larger portion is reflected. As ϕ becomes larger, more and more of the light is reflected, instead of being transmitted. When ϕ reaches a

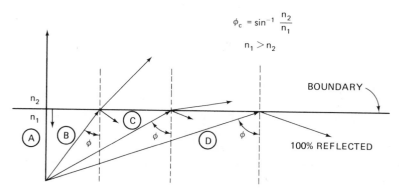

Figure 4.8   Total internal reflection of light rays. For total internal reflection, ϕ must be greater than ϕ_c. (From Ref. 9; copyright 1979 T&B/Ansley.)

certain critical angle ϕ_C, all the light will be reflected. Because of this *total internal reflection,* the light will be confined to the core and will follow a zigzag path (Fig. 4.9) down the core to the other end.

From Snell's law (Chapter 2) we can compute the critical angle by

$$\sin \phi_C = \frac{n_2}{n_1}$$

provided that n_1 is greater than n_2 ($n_1 > n_2$). This can be rewritten

$$\phi = \sin^{-1} \frac{n_2}{n_1}$$

For a typical case, $n_1 = 1.48$, $n_2 = 1.46$, and $\phi_C = 80.6$ degrees. Note that total internal reflection will occur only for those rays incident at angles equal to or greater than the critical angle.

In traveling down the optical fiber, each ray of light may be reflected hundreds or thousands of times. The rays reflected at high angles—the high-order modes—must travel a greater distance than the low-angle rays to reach the end of the fiber. Because of this longer distance, the high-angle rays arrive later than the low-angle rays. As a consequence, modulated light pulses broaden as they travel down the fiber. The output pulses then no longer ex-

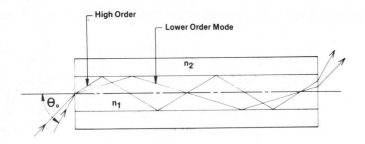

Ray Diagram — Step Index Fiber

Figure 4.9 Refractive confinement in step-index fiber. (From Ref. 9; copyright 1979 T&B/Ansley.)

actly match the input pulses, causing signal distortion. This is discussed in more detail in Section 4.4.

In a graded-index fiber, light rays will travel at different speeds in different parts of the fiber because the refractive index varies throughout the fiber. Near the outer edge, the index is lower; as a result, rays near the outer edge (outer extremity of the core) will travel faster than rays in the center of the core. Because of this higher speed such rays will arrive at the end of the fiber at approximately the same time, even though they took longer paths.

In effect, light rays in these fibers are continually refocused as they travel down the fiber. This refocusing reduces dispersion (to be defined later) and permits operation at much higher data rates.

Light rays strike the end surface of an optical fiber at many different angles. However, for a ray to be propagated down a fiber, it must enter the end of the fiber within a region called the acceptance cone, shown in Fig. 4.10. That is, a light ray not within the cone will get lost in the cladding and never make its way down the core.

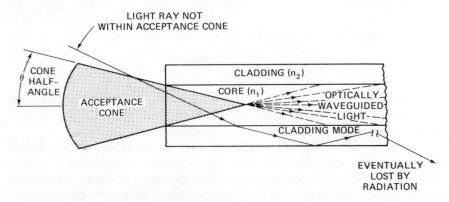

Figure 4.10 Optical fiber's acceptance cone half-angle. (From Ref. 11; copyright 1978 Cahners Publishing Co., *EDN.*)

The half-angle of this cone θ is defined for step-index fibers as

$$\sin \theta = n_1 - n_2$$

(*Note:* $\sin \theta$ is also the numerical aperture, discussed in Section 4.3.) This angle is the maximum angle with respect to the fiber axis at which rays can be accepted for transmission through the fiber. For graded-index fibers, $\sin \theta$ is the same as the numerical aperture (NA).

The graded-index fiber is more difficult to manufacture and is therefore more expensive. It has a better bandwidth than the multimode step-index fiber. It is best for applications requiring long-distance links with low dispersion. Single-mode step-index fibers have the widest bandwidth but are difficult to splice and connect because of their very small diameter. They also require a laser as a light source. Multimode step-index fibers are good for those short distances where a high numerical aperture is needed. LEDs can be used with multimode fibers.

4.2.2 Bandwidth

Optical fibers, as has been noted earlier, have a higher bandwidth than conventional coaxial cable. For coaxial cable the bandwidth varies inversely as the square of the length [12]. For optical fibers, bandwidth is inversely proportional to length for distances up to 1 km. At greater distances the bandwidth is roughly inversely proportional to the square root of the length [13].

Thus, bandwidth figures include a length (km) or assume it: for example, 100 MHz-km. If km is not given, it can be assumed. Any given bandwidth is generally based on a length of 1 km. Fibers with a bandwidth of 3 GHz-km have been developed. Bandwidths of 18 GHz-km for optical fibers are considered possible [13]. Bandwidth is limited by pulse dispersion (discussed in Section 4.5).

4.2.3 Bidirectional

Although most light transmission in fiber optics is in one direction only in any given fiber, bidirectional transmission is possible. This has been accomplished by transmitting two different wavelengths simultaneously in opposite directions through one fiber [14].

4.3 NUMERICAL APERTURE

Numerical aperture is a measure of the light-gathering or collecting power of an optical fiber. The larger the numerical aperture (NA), the greater will be the amount of light accepted by the fiber. Thus, as the NA increases, the greater will be the possible transmission distance, assuming that the same light source and detector are used. Unfortunately, as the NA is increased, the bandwidth is decreased.

As we noted earlier, NA is a function of the refractive indexes of the fiber. It is always less than 1. Typical values are 0.21, 0.30, 0.5, and so on. For a step-index fiber,

$$NA = \sqrt{n_1^2 - n_2^2}$$

For graded-index fibers,

$$NA = \sin \theta$$

where θ is the half-angle of the acceptance cone of the fiber.

The numerical aperture also indicates the efficiency of source-to-fiber coupling, as indicated in Fig. 4.11.

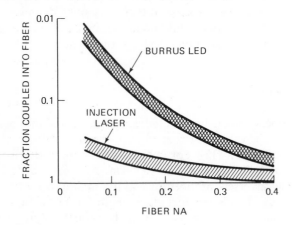

Figure 4.11   Coupling loss vs. fiber numerical aperture. (From Ref. 15; courtesy of General Cable Co.)

4.4 TRANSMISSION LOSSES

The transmission loss or attenuation of an optical fiber is perhaps the most important characteristic of the fiber, as it determines if a system is practical. It dictates (1) spacing between repeaters and (2) the type of optical transmitter and receiver to be used.

As light waves travel down an optical fiber, they lose part of their energy because of various imperfections in the fiber. These losses (or attenuation) are measured in decibels per kilometer (dB/km). For any given cable the attenuation will of course be the fiber attenuation (dB/km) multiplied by the length (km) of the cable. Obviously, the greater the attenuation, the less will be the light that reaches the light detector (receiver).

Typical low-loss fibers have an attenuation of 2 to 4 dB/km for the wavelengths used in fiber optics. Contrast this attenuation with that shown by coaxial cables in Fig. 4.12.

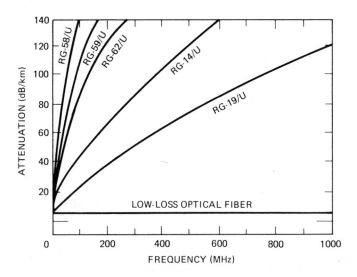

Figure 4.12   Cable attenuation vs. frequency characteristics. (Courtesy of AMP.)

For fibers and coaxial cables alike, the losses are a function of frequency. That is, fibers have greater losses at some frequencies than at others. These losses are usually specified at certain wavelengths rather than at certain frequencies. A typical spectral attenuation plot for a step-index fiber is given in Fig. 4.13.

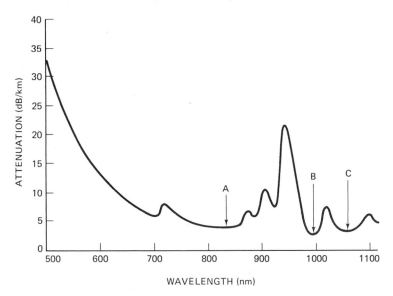

Figure 4.13   Typical fiber attenuation curve: attenuation vs. wavelength, QSF-A grade material; NA, 0.18. (Courtesy of Quartz & Silice.)

Notice the minimum points on the curve at *A, B,* and *C.* These points correspond approximately to attenuations of 4, 3, and 4 dB/km for wavelengths of 830, 1000, and 1060 nm, respectively. If these wavelengths are deviated from, notice how high the losses become.

Figure 4.13 represents a typical fiber. Fibers with lower losses are available commercially, and fibers with very low losses have been developed in the laboratory. The Nippon Telegraph and Telephone Public Corp., for instance, has made a fiber with an attenuation of only 0.20 dB/km at a wavelength of 1550 nm [16].

When suitable light sources and receivers become available for the 1500-to 1600-nm band, this band is likely to be the new standard wavelength, instead of the present 800- to 900-nm band.

Some fibers are being made with double windows. That is, they have minimum attenuation in both the 800- to 900-nm band and the 1500- to 1600-nm band. When such fibers are installed, they can be used with presently available 800- to 900-nm light sources and receivers. Later, when 1500-to 1600-nm sources and receivers become available, the system can be upgraded by simply replacing the light source and detector, leaving the fiber unchanged.

Regardless of wavelength, low-loss fibers are generally the most expensive. For that reason, they are specified only when necessary, usually for long-distance links.

Optical-fiber-measuring techniques have not been standardized, so it is difficult to compare one fiber manufacturer's attenuation figures with another's. In an investigation sponsored by the National Bureau of Standards, seven fiber manufacturers reported measurements of 4.8 to 9.61 dB/km on the same fiber [17].

Attenuation figures are specified for a specific wavelength and numerical aperture (NA), as shown in Table 4.1. Larger NA fibers have higher transmission losses.

Temperature changes may affect attenuation, but even so, optical fibers are more stable than coaxial cable with such changes [18].

Attenuation results primarily from absorption and scattering. Optical radiation losses are less significant as a source for attenuation. All three types of losses are discussed in the following paragraphs.

4.4.1 Absorption

As compared to ordinary glass, optical fibers are remarkably free from impurities. Through careful processing, for instance, silicon has been kept to a purity of 99.9999% [16]. However, even though the impurities are only a few parts per billion, they absorb some of the light and convert it to heat.

TABLE 4.1 Optical-Fiber Characteristics

Fiber Type	Outside Diameter (μm)			Numerical Aperture	Attenuation (dB/km)	Bandwidth (−3dB) MHz-km	Vendor	Cost/m 1 km qty	Min. Bend Radius	Tensile Strength	Remarks
	Core	Cladding	Fiber								
All-plastic (step-index)	368	400		0.53	320 at 690 nm		DuPont	$1.95	1.5	25 kg	Pifax PIR 140 (cables only)
	92	400		0.53	385 at 650 nm		DuPont	$1.45	1.5	25 kg	Pifax P140 (cables only)
Plastic-clad silica (PCS) (step-index)	200	600		0.4	40 at 775 nm		DuPont	$1.95	3	65 kg	Pifax S120 (cables only)
					50 at 820 nm						
	125	300	500	0.3	35 at 790 nm		ITT		5	5×10^5 psi	T301, 2, 3
					20 at 790 nm			$0.50			
					10 at 790 nm			$0.55			
	200	350	500	0.3	35 at 790 nm		ITT		8	5×10^5 psi	T321, 2, 3
					20 at 790 nm			$0.75			
					10 at 790 nm			$0.85			
	250	550		0.3	10 at 800 nm	25	Valtec				PC10
	200	500		0.3	10 at 800 nm	25					PC08
	200	400		0.19	25	30	Thomas & Betts		8	6 kg	Made by Fort Fiber Optics, Paris, France
	400	600									

TABLE 4.1 (*cont.*)

Fiber Type	Outside Diameter (μm)			Numerical Aperture	Attenuation (dB/km)	Bandwidth (−3dB) MHz-km	Vendor	Cost/m 1 km qty	Min. Bend Radius	Tensile Strength	Remarks
	Core	Cladding	Fiber								
All-glass (step-index) multimode	62.5	125	200	0.16	7 at 800 nm / 10 at 800 nm	50	Times		4	10^5 psi	
	100	150		0.3	8 at 830 nm / 15 at 830 nm / 20 at 830 nm	10	NEC				Selfoc SF-S1-100PR 100A 100B
	55	125	500	0.25	12 at 850 nm / 8 at 1060 nm / 8 at 850 nm / 5 at 1060 nm / 5 at 850 nm / 3 at 1060 nm		ITT	$0.65 / $0.95	5	5×10^5 psi	T101, 2, 3
	50 / 100	125 / 150		0.20 / 0.29	10	30	Thomas & Betts		10	2 kg / 1.5 kg	Fort Fiber Optics
All-glass (step-index) single-mode	5	100		0.1	20 at 633 nm / 8 at 800–900 nm / 5 at 1060 nm		Valtec				SHO5-A,B,C
	4.5	80	500	0.1	4–20 at 630 and 850 nm	500+	ITT		5		T-110

TABLE 4.1 (*cont.*)

All-glass multimode	62.5	125	350	0.2	8–3	100–400	Valtec	$0.40–$1.35		5 × 10⁴ psi	13 Models: M605-810, 510, 520,530,540, 410,420, 430, 440,310, 320, 330,340
	55	125	500	0.25	12 at 850 nm 8 at 1060 nm 5 at 1060 nm 8 at 850 nm 5 at 1060 nm 5 at 850 nm 3 at 1060 nm		ITT	$0.60 $0.70 $1.15	5	5 × 10⁵ psi	T201, 2, 3
	55	125	500	0.25	Same as T201,23	400	ITT	$0.75–$1.50			T211, 2, 3, T-22, 1, 2, 3
	60	150		0.2	8 at 830 nm 10 at 830 nm 15 at 830 nm	300 200	NEC				SF-G1-GOPR SF-G1-60A SF-G1-60B
	62.5 85–90	125 147	200 400	0.16 0.22	5–10 at 800 nm 5 at 840 nm	200–600 200	Times Northern Telecom		4 3	10⁵ psi 50 newtons	
	100	140	400	0.30	7 at 820 nm	20 at 900 nm	Corning	$0.70		10⁵ psi	Short-distance fiber (SDF) or "Fat Fiber"
	200	230		0.4	35 at 820 nm	5 at 900 nm	Corning		15		"Super Fat Fiber"

TABLE 4.1 (*cont.*)

| Fiber Type | Outside Diameter (μm) | | | Numerical Aperture | Attenuation (dB/km) | Bandwidth (−3dB) MHz-km | Vendor | Cost/m 1 km qty | Min. Bend Radius | Tensile Strength | Remarks |
	Core	Cladding	Fiber								
All-glass multimode		125	138	0.21	4–8 at 820 nm	200–500 at 900 nm	Corning	$0.45–$2.50		2.5 × 10⁴ psi	14 models: 4150, 4100, 4080, 4040, 5100, 5080, 5040, 5020, 6100, 6080, 6040, 6020, 8040, 8020
	63	125	138	0.21	2–5 at 900 nm Less than 2 at 1060 nm and 1300 nm also available	200–1000	Corning	$0.70–$1.75		2.5 × 10⁴ psi	Long-wavelength series 11 models: 2041, 3101, 3081, 3041, 3021, 4101, 4081, 4041, 5081, 5041, 5021
	50	125		0.20	6	300	Thomas & Betts		10	2 kg	Fort Fiber Optics

Source: Reprinted with permission from Ref. 18 (*Electronic Design*, Vol. 27, No. 23); copyright Hayden Publishing Co., Inc., 1979.

4.4.2 Scattering

Variations in the molecular density and composition of the fiber cause scattering of the light. This scattering limits the shorter wavelengths. The loss is inversely proportional to the fourth power of the wavelength.

4.4.3 Radiation

During fiber-cable manufacture and installation, very small but sharp bends sometimes occur accidentally in the fibers. These random bends (or axial distortions) are shown in Fig. 4.14. Light can radiate or escape at these microbends, causing transmission loss. With recent improvements in cable design, however, this loss has become less significant.

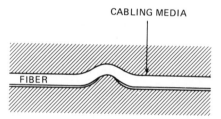

Figure 4.14 Microbending due to cabling process. (From Ref. 9; copyright 1979 T&B/Ansley.)

4.5 SIGNAL (DELAY) DISTORTION

As light pulses travel along an optical fiber, they tend to become wider, often to the point that they will overlap other pulses and smear the information (see Fig. 4.15). This pulse dispersion (or spreading or broadening) makes it hard for the receiver to tell one pulse from another. It is a form of signal distortion which effectively limits the information-carrying capacity of a fiber-optic system.

Pulse dispersion is primarily a result of modal and material dispersion, which are discussed in the following paragraphs.

4.5.1 Modal Dispersion (Also Called Intermodal or Multimode Dispersion)

Modal dispersion can be reduced by using a multimode graded-index optical fiber. It can be substantially eliminated by using a single-mode step-index fiber. The single-mode fiber, however, has some disadvantages as we discussed earlier.

The difference in width (in nanoseconds) between an input pulse and its corresponding output pulse is the pulse dispersion. As it is related to distance, it is generally specified in nanoseconds/kilometer (ns/km). For sufficiently long fibers it is proportional to the square root of the fiber length [11].

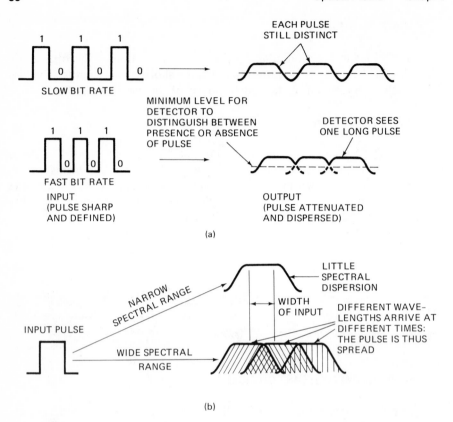

Figure 4.15 Pulse dispersion: (a) modal dispersion; (b) material dispersion. (From Ref. 10; courtesy of AMP.)

In a step-index fiber, light rays that travel parallel to the axis will have a shorter path length than rays that zigzag down the fiber, as shown in Fig. 4.16. Consequently, some rays will take longer to reach the output. For a 1-km fiber with a core refractive index (n_1) of 1.48, a cladding index (n_2) of 1.46, and a core diameter of 50 μm, Kleekamp and Metcalf [11] have

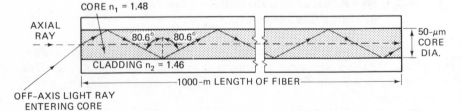

Figure 4.16 Time delay in step-index fiber. *Note:* The off-axis light ray follows a zigzag path 1014 m long, compared to 1000 m for the axial ray. The extra 14 m produces an arrival-time difference of 69 ns. (From Ref. 11; copyright 1978 Cahners Publishing Co., *EDN*.)

calculated that an off-axis light ray incident at an angle of 80.6 degrees would travel 1014 m to reach the end of a fiber 1000 m long. An axial ray in the same fiber would travel only 1000 m. The difference of 14 (1014 minus 1000) means that the off-axis ray would arrive later than the axial ray, even if both rays started at the same instant.

Just how much later?

As we have seen earlier,

$$v = \frac{c}{n_1}$$

Therefore,

$$v = \frac{3 \times 10^8 \text{ m/s}}{1.48}$$

$$= 2.03 \times 10^8 \text{ m/s}$$

Since time equals distance divided by velocity, it will take 69 nanoseconds (14 m divided by 2.03×10^8) longer for the off-axis ray to arrive.

This time delay is a measure of the pulse dispersion. It can be as low as 0.3 ns/km. If this delay is comparable to the interval between pulses, the output pulses will overlap or spread into adjacent time slots. As a result, the receiver will no longer be able to determine what was sent. In the preceding example, if pulses occurred more frequently than every 69 ns, they would be indistinguishable.

4.5.2 Material (Chromatic) Dispersion

With a true monochromatic (one color) light source, there is no material dispersion. However, the optical signal of an LED in particular and in the injection laser to a slight extent is a combination of many frequencies. Instead of emitting just one wavelength, these light sources emit a band of wavelengths.

As we noted in Chapter 2, different wavelengths have different velocities in glass. All portions of the input light pulse, which is a combination of wavelengths, will not arrive simultaneously at the output.

The result is a distortion of the optical signal. This distortion is not a problem now but is expected to be in future wide-bandwidth systems. It may not be a problem even then if a wavelength of 1200 to 1600 nm is used, as material dispersion is much less at this wavelength.

By replacing an LED (wide spectral width) with an injection laser (narrow spectral width), dispersion can be reduced by a factor of 20 [19]. Material dispersion is a problem with graded-index fiber but can be ignored with step-index fibers [20].

In a typical case, the material dispersion for an injection laser transmitting at 820 nm with a 4-nm spectral width is about 0.4 ns in 1 km [21].

4.6 CABLE CONFIGURATION

A fiber-optic cable is one or more optical fibers formed into a cable for convenience and protection. Whether it is buried directly in the ground, hung on telephone poles, pulled through underground ducts, or dropped to the bottom of a lake or ocean, this cable is likely to receive much abuse and mistreatment during its lifetime.

While being installed, it may be stepped on and banged about. Trucks and drums may roll over it. As it is being pulled through ducts, it may be stressed beyond expectation. Once in place it may be subjected to a very cold Canadian winter or a hot Nevada summer. Ice may load it, causing it to sag or break. Gophers and other rodents may try to chew through it. In ice-clogged ducts, technicians may hit it with steam as they clear the ducts. It may be submerged in water in flooded manholes.

The cable must be able to survive this abuse, yet it must be reasonably easy to repair if it breaks, be economically competitive with conventional cables, and be space efficient.

Numerous designs or configurations have been developed to meet these requirements as indicated in Figs. 4.17 and 4.18 and Table 4.2. These designs differ in materials and arrangements, but practically all of them include coatings to protect individual fibers, strength-bearing materials, filler or buffer materials, and an external protective jacket. In addition, for specific applications some cables include armor protection against rodents and copper wires for carrying electrical power.

The optical fiber is coated with soft silicone immediately upon fabrication to prevent damage from abrasion and moisture. This coating is considered part of the fiber, not the cable. An additional coating or jacket of a durable plastic may be added for still further protection. The jacket may be color-coated for easy identification during installation and repair.

The strength or tension member minimizes or eliminates stretching force (tensile stress) applied to the optical fibers. It is also called the load-bearing member. It is made of steel or braided Kevlar aramid yarn. As Kevlar is nonconducting, it is better than steel in places where the cable would be susceptible to lightning damage. Polyurethane cable fillers or buffers help to cushion the fibers.

The outer protective jacket may be made of polyethylene, polyurethane, polyvinyl chloride (PVC), or Tefzel. It protects the fibers from dirt, moisture, sunlight, abrasion, crushing, and temperature variations. Like the individual fibers, it may also be color-coded. Length markers and cable type may be imprinted on this jacket. Flame-retardant types are available. In some cases this jacket may carry some of the load, just as the strength members do.

As shown in Fig. 4.19, fibers may be placed in plastic tubes (buffers) before being placed in a cable. If the fiber has a *loose* fit in the tube, it can move freely inside and thereby theoretically provide protection against stress, as shown in Fig. 4.20. In the *tight*-fit case, the tube is filled with polyurethane

TABLE 4.2 Typical Cable Parameters[a]

Belden Trade No.	No. of Fibers	Overall Dia. (mm)	Buffer Tube I.D. (mm)	Buffer Tube O.D. (mm)	Weight (kg/km)	Recommended Max. Installation Pulling Strength (lb)	Bending Radius (mm)
227001	1	3.8	1.2	2	13.5	260	50
227002	2	3.8 × 7.6	1.2	2	27	520	50
227006	6	8	1.2	2	40	480	100
227012	12	14	1.2	2	155	800	150
227018	18	18	1.2	2	260	800	200

[a] All dimensions are nominal. Standard lengths 1 km.

Source: Courtesy of Belden Corporation.

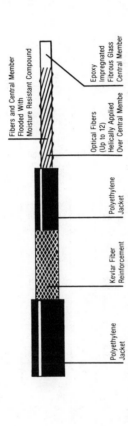

(a) TYPE NM

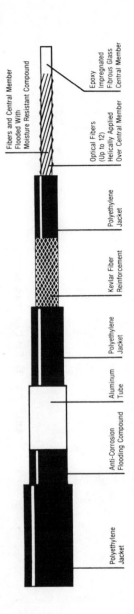

(b) TYPE AT

(c) TYPE FPA

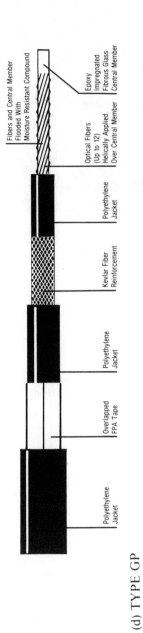

Fibers and Central Member Flooded With Moisture Resistant Compound

Epoxy Impregnated Fibrous Glass Central Member

Optical Fibers (Up to 12) Helically Applied Over Central Member

Polyethylene Jacket

Kevlar Fiber Reinforcement

Polyethylene Jacket

Overlapped FPA Tape

Polyethylene Jacket

(d) TYPE GP

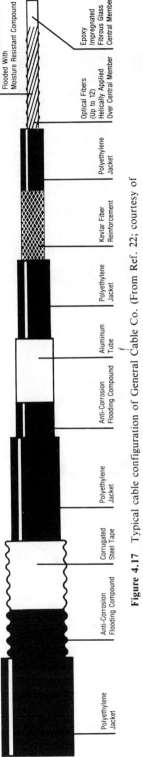

Fibers and Central Member Flooded With Moisture Resistant Compound

Epoxy Impregnated Fibrous Glass Central Member

Optical Fibers (Up to 12) Helically Applied Over Central Member

Polyethylene Jacket

Kevlar Fiber Reinforcement

Polyethylene Jacket

Aluminum Tube

Anti-Corrosion Flooding Compound

Polyethylene Jacket

Corrugated Steel Tape

Anti-Corrosion Flooding Compound

Polyethylene Jacket

Figure 4.17 Typical cable configuration of General Cable Co. (From Ref. 22; courtesy of General Cable Co.)

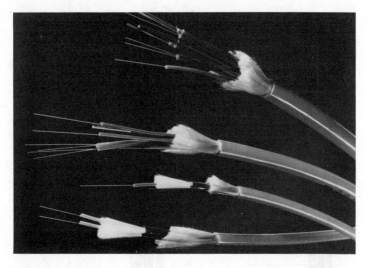

Figure 4.18 Typical cable configurations of Siecor. (Courtesy of Siecor.)

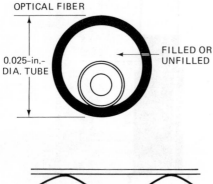

Figure 4.19 Optical fiber in buffer tube. (From Ref. 15; courtesy of General Cable Co.)

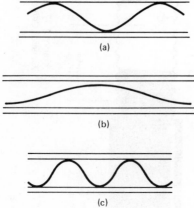

Figure 4.20 Optical fiber with excess length inside a loose buffer jacket: (a) fiber in buffer jacket after cable manufacturing; (b) decrease of fiber excess length caused by strain of buffer jacket during cable stress; (c) increase of fiber excess length caused by shrinkage of buffer jacket materials during cooling. (From Ref. 23; courtesy of Siecor Optical Cable and Siemens AG.)

and the fiber is encased. Loose-fit fibers are represented by a dot in a circle in cable cross-section drawings.

Although loose-fit or bound fibers have been considered better than tight-fit fibers for most uses, as least one test has contradicted this. In a demonstration system built at the Mitre Corporation for the Air Force, tightly bound fibers did not break after 30,000 vehicles had run over them. But in the loosely bound fibers, breaks began to occur after 500 traffic counts [24].

In the early days of fiber optics, many high-loss fibers were grouped in a bundle to form an effective single low-loss fiber. But now the bundle concept is being abandoned because single fibers are much smaller and more effective.

Most often, these fibers are used in pairs to allow for duplex operation. When bidirectional fiber-optic systems become common, a single fiber can be used for full duplex operation. Two fibers may be assembled into a cable, but generally a cable will include more than two. In fact, optical-fiber cables may include a dozen or so single fibers.

The single fibers may be placed in ribbons as in Figs. 4.21 and 4.22, or in circularly symmetric units as shown in Figs. 4.23 and 4.24. The advantages and disadvantages of these two basic types are given in Tables 4.3 and 4.4.

Cable design involves several trade-offs. Improvements in some characteristics may cause losses in others. Obviously, thicker coatings give greater moisture and abrasion protection, but they make the cable bulkier and therefore harder to install. In addition they take up too much space in already overcrowded ducts.

In constructing these cables, the goal is to make them rugged enough to withstand the same environments as are withstood by coaxial cables or twisted-pair cables. That is, fiber-optic cables must not require pampering or white-glove treatment.

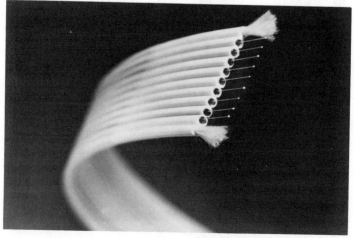

Figure 4.21   Ribbon cable. (Photo courtesy of Siecor.)

RIBBON

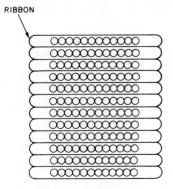

Figure 4.22 Stranded rectangular array ribbon cable. (From Ref. 25; copyright 1979 Bell Telephone Laboratories, Inc.; reprinted by permission.)

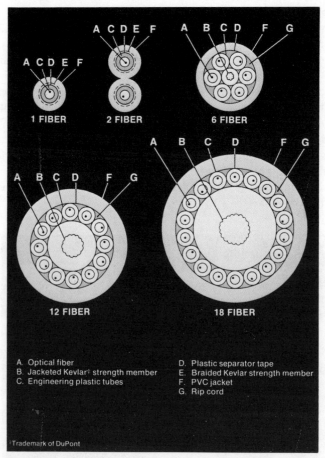

Figure 4.23 Typical cable configurations with loose fit in buffer tubes. (Courtesy of Belden Corporation.)

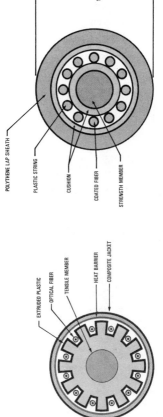

POLYTHENE LAP SHEATH

PLASTIC STRING

CUSHION

COATED FIBER

STRENGTH MEMBER

18 mm

EXTRUDED PLASTIC

OPTICAL FIBER

TENSILE MEMBER

HEAT BARRIER

COMPOSITE JACKET

(A) - Optical cable design in which the fiber runs in a void. Cables of this type have a relatively large diameter and the fiber must support its own weight.

(B) - Optical cable design in which the fiber is supported by soft cushions. This is similar to the previous design but the fiber has less freedom of movement.

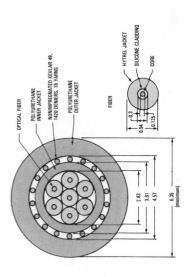

OPTICAL FIBER

POLYURETHANE INNER JACKET

NONIMPREGNATED KEVLAR 49, 1420 DENIERS, 18 YARNS

POLYURETHANE OUTER JACKET

2.82

3.81

4.57

6.35 (maximum)

FIBER

HYTREL JACKET

SILICONE CLADDING

CORE

0.3

0.94

0.175

(C) - Optical cable incorporating fully supported fibers. This design has a small cross section, high positional stability, and good impact resistance. However it does require fiber with a low sensitivity to bending.

Figure 4.24 Three main categories of optical cable design. (From Ref. 26; reproduced by permission of Electrical Communication.)

97

TABLE 4.3 Comparison of Cable Design Approaches

Factor	Circularly Symmetric Unit	Flat Ribbon Unit
Microbending loss	Easier to control	Harder to control but doable
Loss temperature stability	Easier to control	Harder to control but doable
Space efficiency (fibers/area)	Lower	Higher
Ease of fiber handling and identification	More complex	Simpler
Suitability for mass splicing	Poorer	Superior
Suitability for repair splicing	Poorer	Superior
Ease of manufacture	Good	More difficult
Compatibility with conventional cable manufacture	Good	Poor

Source: Ref. 25, p. 452; copyright 1979 Bell Telephone Laboratories, Inc,; reprinted by permission.

TABLE 4.4 Bell Labs' Preferred Designs
for Communication Applications

Application	Bandwidth (Mbit/s)	Repeater Spacing (km)	No. of Fibers—M	Preferred Cable Structure[a]
Point to point—either long haul or interoffice	< 100	10 or less	> 10	FRU
	> 100		< 10	CSU
		Greater than 20		CSU
Large feeder—with branching capability			> 100	FRU
Small feeder or distribution			10 < M < 30	?
			50 <	FRU
	< 20		30 <	FRU
Drop cables			< 4	CSU

Source: Ref. 25, p. 453; copyright 1979 Bell Telephone Laboratories, Inc.; reprinted by permission.

[a] FRU, flat ribbon unit; CSU, circular symmetric unit.

REFERENCES

1. Fujitsu product literature, n.d.
2. John Free, "Fiber Optics Start a Revolution in Our Telephone Systems as Threads of Light Replace Copper Wires," *Popular Science,* May 1980, pp. 99–103, 180.

3. Les Borsuk, "Introduction to Fiber Optics—What All Connector Engineers Need to Know," ITT Cannon Electric, Santa Ana, Calif., n.d.

4. ITT product sheets, 1978.

5. "Standard Fiber Dimensions Are Proposed by IEC; Approval Is Expected This Year," *Laser Focus,* Feb. 1980, pp. 84–86.

6. "Military Communications," *Laser Focus,* Nov. 1979, p. 101.

7. "Design Criteria for Optical Communication Systems," Galileo Electro-Optics Corp., Technical Memorandum 200A, Mar. 1979.

8. Lee L. Blyler, Jr., and Shiro Matsuoka, "Polymer Protection for Glass Fibers," *Bell Laboratories RECORD,* Dec. 1979, pp. 315–319.

9. "Application Notes, Introduction to Fiber Optics," T&B /Ansley Publication No. AFO-1000, 1979. (Drawings reproduced with permission from Thomas & Betts Corp.)

10. "Introduction to Fiber Optics and AMP Fiber-Optic Products," HB 5444, AMP Incorporated, n.d.

11. Charles Kleekamp and Bruce Metcalf, "Designer's Guide to Fiber Optics—Part 1," *EDN,* Jan. 5, 1978.

12. G. S. Anderson, J. C. McNaughton, and R. L. Ohihaber, "Fiber Optic Cables for Telecommunications," Third International Telecommunication Exposition, Dallas, Tex., 1979.

13. John Kane, "Fiber Optic Cables Compete with MW Relays and Coax," *Microwave Journal,* Jan. 1979, pp. 16, 61.

14. "New Products Abound at Exhibition, But Few New Firms Display Hardware," *Laser Focus,* Nov. 1979, p. 56.

15. Herb Lubars, "Optical Fiber Cable Systems," presented at the Western Regional Meeting, AAR, May 1, 1979.

16. "News Briefs: NTT Drops Fiber Loss Again," *Electronics,* Mar. 29, 1979, p. 48.

17. "Optical Fiber Specs: To Believe or Not to Believe?" *Microwaves,* Sept. 1979, p. 18.

18. Stephen Ohr and Sid Adlerstein, "Fiber Optics Is Growing Strong," *Electronic Design,* Vol. 27, No. 23, Nov. 8, 1979.

19. Enrique A. T. Marcatili, "Objectives of Early Fibers: Evolution of Fiber Types," in *Optical Fiber Telecommunications,* ed. Stewart E. Miller and Alan G. Chynoweth (New York: Academic Press, Inc., 1979), p. 28.

20. Belden Fiber Optic Technical Bulletin.

21. Alan F. Fairaizl, "How to Select Fiber-Optic Cables for Practical Applications," Siecor, n. d.

22. G. H. Foot and J. B. Masterson, "Optical Fiber Cable Installation Experience," presented at the Second U.S./Southeast Asia Telecommunications Conference, Dec. 3, 1980.

23. Peter R. Bark, Ulrich Oestreich, and Gunter Zeidler, "Fiber Optic Cable Design, Testing and Installation Experiences," Siecor Optical Cables, Inc., Horseheads, NY, and Siemens AE, Munich, W. Germany, Nov. 1978.

24. C. W. Kleekamp and B. D. Metcalf, "Fiber Optics for Tactical Communications,"

Air Force Electronic Systems Division Report ESD-TR-79-121, Mitre Corp., Bedford, Mass., MTR-3723, Apr. 1979, p. 34.

25. Morton I. Schwartz, Detlef Gloge, and Raymond A. Kempf, "Optical Cable Design," in *Optical Fiber Telecommunications* ed. Stewart E. Miller and Alan G. Chynoweth (New York, Academic Press, Inc., 1979), p. 441.

26. C. K. Kao, "Optical Fiber Communications Technology," *Electrical Communication,* Vol. 54, No. 3, 1979.

5

SPLICES, CONNECTORS, AND COUPLERS

In fiber-optic systems major light losses can occur at three optical junctions:

1. From source to fiber
2. From fiber to fiber
3. From fiber to photodetector

Whether these junctions are permanent splices or demountable connectors, considerable care must be taken to keep the losses to a minimum.

Junction losses are most often the result of the following problems [1], as shown in Fig. 5.1:

1. *Axial or lateral misalignment* (*displacement*). The fiber axes should be aligned to within 5% of the smaller fiber diameter.
2. *End separation* (*gap misalignment*). In splices the fibers should normally touch, as the farther apart the fibers are, the greater will be the loss. In connectors, however, the ends are intentionally separated so that they will not rub against and damage one another during engagement (mating).
3. *Angular misalignment* (displacement). Ideally, the ends of the fibers should be parallel, as any angle will introduce loss. Any misalignment should be held to under 2 degrees.
4. *Surface finish*. The ends need to be smooth and perpendicular (square). If the fiber faces are more than 3 degrees off perpendicular, losses will exceed 0.5 dB [2].

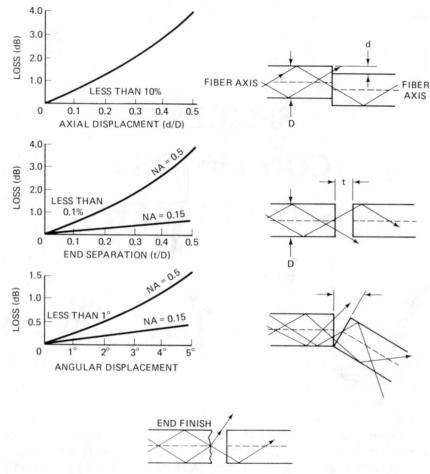

Figure 5.1 Connector-loss considerations. (From Ref. 1; courtesy of AMP.)

To minimize these problems, it is necessary to use special connectors and splicing devices. In addition, special tools must be used to cut, align, and hold the fibers while the splice is being made or the connector is being attached.

The first step in splicing two fibers or placing a connector on a fiber is to remove the outer plastic jacket of the fiber-optic cable with a single-edge razor blade or with any of several special tools made for this purpose. The Kevlar strength members should be pulled back but not cut. Once all protective jackets are removed, the fiber is then wiped clean with isopropyl alcohol, Freon TF, or whatever substance is recommended by the fiber manufacturer. About 1 in. of the fiber needs to be exposed. When exposed, the fiber is extremely fragile and therefore must be handled very carefully.

Once the fiber is exposed, the technician is ready to proceed with either splicing or connectoring, as described in the following paragraphs.

5.1 SPLICING OPTICAL FIBERS

In the world of copper conductors, splicing can be as simple as twisting two wires together and soldering them. Splicing optical fibers, however, is a much more complicated task. Special training, practice, and equipment together with patience and good coordination are necessary before the technician can make acceptable splices.

Proper splicing is difficult for two principal reasons: (1) the hairlike optical fibers are so fine they are hard to handle, and (2) the two fibers must be precisely aligned to keep losses to an acceptable level.

The two basic methods of splicing are fusion and mechanical splicing. In the fusion technique the two fiber ends are placed between electrodes that cause the tips of the fibers to melt and then fuse. In the mechanical method the fibers are clamped together, and then, most often, a transparent epoxy is used to glue the ends together.

Fusion splicing equipment costs more than mechanical splicing equipment, but some believe that it gives a stronger splice [3] and lower losses (generally less than 0.5 dB) [4]. In a recent survey [4] of operational fiber optic systems, six used epoxy (mechanical) splicing, four fusion, and five a combination of fusion and epoxy.

An important factor in the selection of a splicing method is the location where the splicing will be done. Splicing is much easier in a relatively clean, no-wind factory or laboratory than in a dirty outside manhole where a 30-mph wind may be blowing.

A common technique for preparing the fiber end for splicing is the scribe-and-break method (also called score-and-break). In this method the stripped fiber is put under tension and then a sapphire- or diamond-edge cutting tool (Table 5.1 and Figs. 5.2 to 5.4) is used to nick the fiber. Then the fiber is pulled until it snaps. The result should be a smooth fiber end face.

**TABLE 5.1 ITT's List of Supplies
for Fiber-End Preparation Operations**

Diamond-edged cutter
Epoxy
600-grit wet/dry gum-back polishing paper
Al_2O_3 polishing paste
Gum-back polishing paper (napped)
Optical connector ferrule with jewel
Rubber tubing
Syringe
Isopropyl alcohol
Freon TE or TF (preferred)
Wiping cloths
Heat probe
Wire strippers

Source: Ref. 5; courtesy of ITT.

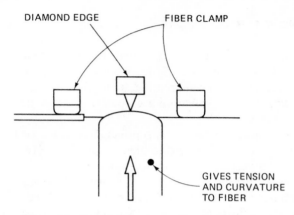

Figure 5.2   Optical-fiber cutter. (Courtesy of Fujitsu.)

Some believe that the scribe-and-break tools which bend the fiber around the circumference of an imaginary circle as the fiber is broken do not produce surfaces as good as those made by the straight-pull type of scribe-and-break tools [2].

5.1.1 Mechanical Splices

Single fibers can be held in place with snug-fitting (precision-machined) tubes (Fig. 5.5), loose-tube splice using square capillary (Fig. 5.6), sandwich splice (Fig. 5.7), or precision pins (Fig. 5.8).

Snug-fitting tubes have had tolerance problems that have discouraged their use.

In the loose tube splice, epoxy is inserted into the tube and then the fibers are forced into the tube.

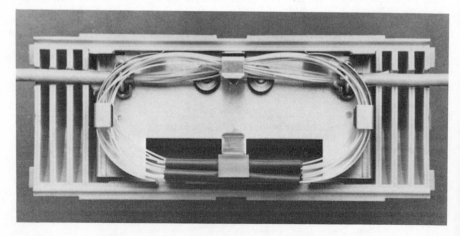

Figure 5.3   Siecor splicing device. (Courtesy of Siecor.)

Figure 5.4 Siecor Model 60 mechanical splicer. (Courtesy of Siecor.)

Figure 5.5 Snug tube splice. (From Ref. 6; copyright 1978 IEEE.)

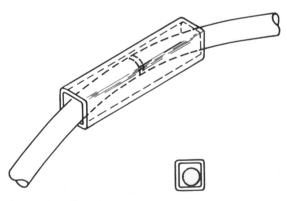

Figure 5.6 Loose tube splice using square capillary. (From Ref. 7; copyright 1979 Bell Telephone Laboratories; reprinted by permission.)

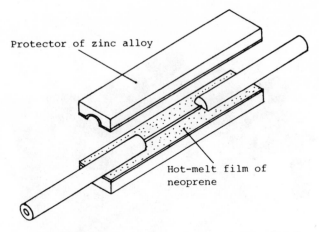

Protector of zinc alloy

Hot-melt film of neoprene

Figure 5.7 Sandwich splice protector (after Arai et al.). (From Ref. 8; copyright 1980 Advanced Technology Publications.)

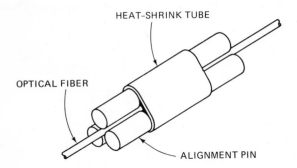

HEAT-SHRINK TUBE

OPTICAL FIBER

ALIGNMENT PIN

Figure 5.8 Precision pin splice. (From Ref. 6; copyright 1978 IEEE.)

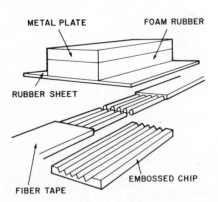

METAL PLATE   FOAM RUBBER

RUBBER SHEET

EMBOSSED CHIP

FIBER TAPE

Figure 5.9 Ribbon splicing. (From Ref. 7; copyright 1979 Bell Telephone Laboratories; reprinted by permission.)

In the precision pin splice the fibers are inserted between three steel pins. An inspection port allows the alignment of the fibers to be checked by means of a microscope. Heat is then applied to the heat-shrinkable sleeving. As the sleeving is heated, it contracts and squeezes the pins together around the fibers. An index-matched epoxy is then applied between the fiber ends. An additional heat-shrink sleeve is then added for further protection.

Whenever epoxies are used in fiber-optic connectors, curing time can

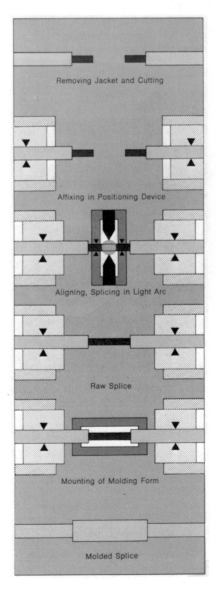

Figure 5.10 Electric arc fusion. (Courtesy of AEG Telefunken.)

often be reduced by application of heat as with an ordinary hair dryer. Some epoxies cure faster when exposed to ultraviolet light.

Multiple fibers formed in a ribbon can be spliced with the groove alignment technique shown in Fig. 5.9. An index-matching material is added to the joint and the cover is then attached.

5.1.2 Fusion Splicing

In fusion splicing the tips of the two fibers are heated until they melt together. Although flames have been used for this procedure, electric arc fusion (Fig. 5.10) is more common.

Figure 5.11 Fusion splicer. (Courtesy of Siecor.)

TABLE 5.2   Materials and Accessories Required for Siecor Model 66 Fusion Splicer

12-volt battery, such as a wet-cell motorcycle battery or Gould Gel-Cell
Protective tubes, two needed per fiber pair to be spliced (furnished with Siecor splice closure)
Solvent for removal of filling compound when splicing filled cables: lighter fluid (e.g., Ronsonol Lighter Fuel)
Acetone for dissolving fibers' lacquer coating
Eyedropper
Metal splice reinforcing parts, one needed per fiber pair to be spliced
Siecor crimping pliers (E-78110) for crimping the metal reinforcing parts
Lacquer for recoating spliced fibers
Plastic toothpicks

Source: Courtesy of Siecor.

TABLE 5.3 Siecor Model 66 Fusion Splicer Specifications

Arc ignition voltage	7.5 kV
Arc sustaining voltage	500 V
Arc current, maximum	25 mA
Maximum fusion temperature	2000°C
Electrodes	
Material	Tungsten
Separation	1.5 mm
Diameter	1.6 mm
Anode length	16 mm
Anode shape	Pointed
Cathode length	55 mm
Cathode shape	Pointed
Lifetime before cleaning	Approx. 100 splices
Power requirements	12 V dc, 1.5 A max.
Fuse	6 A, fast-acting type 3AG
Dimensions	27 cm wide, 29 cm deep, 43 cm high set up, 27 cm high closed
Weight	12 kg

Source: Courtesy of Siecor.

A portable electric arc fusion splicer is shown in Fig. 5.11. It consists of fiber-holding and alignment devices, a viewing microscope, arc positioning and intensity controls, a built-in arc power supply, and a fiber-cutting tool. Materials and accessories for the splicer are given in Table 5.2. Specifications are listed in Table 5.3.

After the fibers are cleaned and cut, they are clamped in the holders of the splicer and positioned with the aid of the microscope. A tiny well-regulated electric arc is used briefly to slightly round the fibers' ends. The fibers are then fused together. Splice loss should be monitored during the operation with an optical time-domain reflectometer (see Section 8.3). After splicing, the joint is recoated and placed in a protective sleeve before placement in a splice closure.

With this splicer, typically splice losses of less than 0.4 dB have been obtained between telecommunications-grade optical fibers. Some of these installations have had splicing losses of only 0.17 dB.

5.2 CONNECTORS

As each connector in a fiber-optic system introduces a loss, proper choice of connectors can make a significant difference in total system losses. Connector losses now range from 0.5 to 2 dB. In terms of percentage power loss,

Figure 5.12 Typical fiber-optic connectors. [(a), courtesy of TRW Cinch; (b) and (c), courtesy of ITT Cannon.]

$$0.50 \text{ dB} = 10\% \text{ loss}$$
$$1.00 \text{ dB} = 20\% \text{ loss}$$
$$1.55 \text{ dB} = 30\% \text{ loss}$$
$$2.22 \text{ dB} = 40\% \text{ loss}$$

The ideal connector must be able to maintain a low insertion loss after repeated mating and unmating. At the same time, it must keep the junction free of contaminants despite hostile mechanical and environmental conditions. And, of course, it needs to be simple and inexpensive. Although the development of such a connector has lagged behind most other fiber-optic components, numerous practical connectors (Fig. 5.12) are now available at reasonable prices. The latest connectors even accommodate power, signal, coaxial, and fiber-optic cables in the same assembly.

From the outside, fiber-optic connectors look much like common electrical connectors. However, their internal mechanisms are more complicated and demand closer manufacturing tolerances.

In the following discussion the emphasis is on connecting two single fibers. Multifiber connectors now available incorporate the same techniques.

Two basic types of connectors are now available: lens-coupled and butt-coupled (the most common type).

In the lens-coupled connector shown in Fig. 5.13, spring-loaded connectors press the fiber ends into low-loss optical lens. The lens are immersed in a fluid whose refractive index matches the fiber core's index. In effect, the fibers are aligned optically rather than mechanically. Thus, it is not necessary to grind or polish the fiber ends for this connection.

Butt-coupled connectors are available in the following varieties:

1. Tube
2. Straight-sleeve
3. Double eccentric

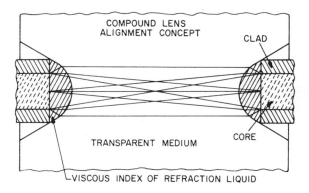

Figure 5.13 Liquid-lens connector developed by the Deutsch Company. (From Ref. 9; copyright 1979 Milton S. Kiver Publications, Inc.)

4. Tapered sleeve
5. Three-rod
6. Jewel bushing
7. Four-pin
8. Resilient ferrules

Tube method (Fig. 5.14).*   In the tube method a metal jack and plug are held together by a threaded coupling (not shown in the figure). The fit of the plug into the jack provides the initial alignment. The fiber in the jack then enters the tapered alignment hole of the plug. Since the alignment hole must be large enough to accept the largest fiber, fiber variations result in axial misalignment. The fibers must also be positioned for the correct end separation with reference to the stops on the jack and plug. During the engagement any contamination in the alignment hole will be pushed forward by the fiber and end up between the fibers.

Straight-sleeve method (Fig. 5.15).   In the straight-sleeve method a precision sleeve is used to mate two plugs, which are often designed after SMA coaxial connectors. The sleeve aligns the fibers. The machining of the

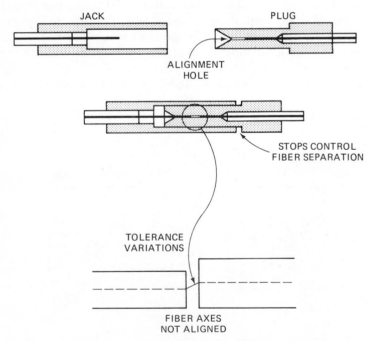

Figure 5.14   Tube-alignment connector. (From Ref. 1; courtesy of AMP.)

* The following description of butt-coupled connectors is taken from Ref. 1.

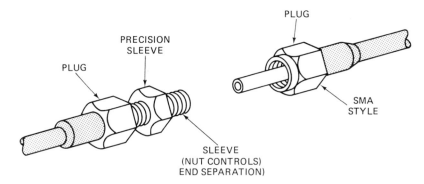

Figure 5.15   Straight-sleeve connector. (From Ref. 1; courtesy of AMP.)

connector is difficult since the fiber holes must be dead-center. Several close tolerances must be met for correct hole placement. Any rotation of mated connectors will affect the loss if the two plugs and the sleeve are not concentric. These connectors are easy to use, but high-quality ones are expensive.

Double-eccentric method (Fig. 5.16). Rather than trying to make a straight sleeve as concentric as possible, some manufacturers use a double eccentric that purposely puts everything off-center. The fibers are held

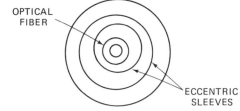

Figure 5.16   Double-eccentric connector. (From Ref. 1; courtesy of AMP.)

within two eccentrics. The two connector halves are mated. The eccentrics can than be rotated to bring the fibers axes into very close alignment, where the lowest losses are had. The eccentrics are then locked in place. Tolerances are reasonable, and performance is good, although the adjustment itself can be cumbersome. The coupling must usually be hooked into test equipment that will indicate the peak adjustment.

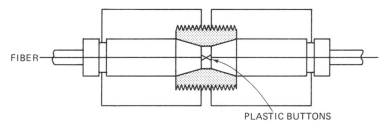

Figure 5.17   Tapered-sleeve method. (From Ref. 1; courtesy of AMP.)

Tapered-sleeve method (Fig. 5.17). The tapered-sleeve method uses a tapered plug and sleeve. Compliant plastic buttons on the ends of the plugs squeeze together and provide matching of optical properties. The fibers are molded into the plastic plugs during manufacture, which is a very precise process. Different variations in fibers are individually matched to a mold to give good performance.

Three-rod method (Fig. 5.18). Three rods may be placed together so that their center space is the size of the fiber. With this arrangement, there is good concentric alignment. The rods, which have equal diameter, compress and center the fibers radially. The rods usually have some compliancy to absorb fiber variations and provide axial alignment. By using the two rows of rods, this method can be adapted to ribbon cable and other multiple-fiber arrangements. Individual rods may be used, or they may be formed as a single unit. An important part of this connector is that the two mating members overlap, allowing both members to compress each fiber.

Jewel bushing (Fig. 5.19) [10]. Readily available watch jewels can be used for precise lateral alignment. Like pistons in a cylinder, two opposing jewel/ferrules are mated in a guide sleeve.

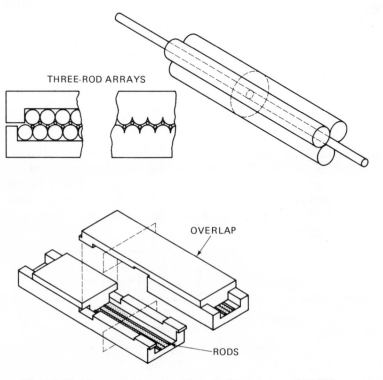

Figure 5.18   Three-rod connector. (From Ref. 1; courtesy of AMP.)

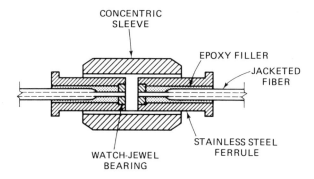

CONCENTRIC
SLEEVE

EPOXY FILLER

JACKETED
FIBER

STAINLESS STEEL
FERRULE

WATCH-JEWEL
BEARING

Figure 5.19   Watch-jewel-bearing connector designed by ITT Cannon. (From Ref. 12; copyright 1978 Milton S. Kiver Publications, Inc.)

Four-pin method (Fig. 5.20).   Four pins held with a ferrule can be used to center the fiber. When used with the straight-sleeve method, the pins are used to center the fiber in the connector, and the sleeve is used to align mating connectors.

TRW Cinch's Optalign (Fig. 5.21) connector [11] is a four-rod glass ar-

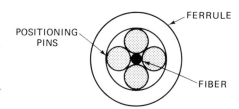

FERRULE

POSITIONING
PINS

FIBER

Figure 5.20   Four-pin method connector. (From Ref. 1; courtesy of AMP.)

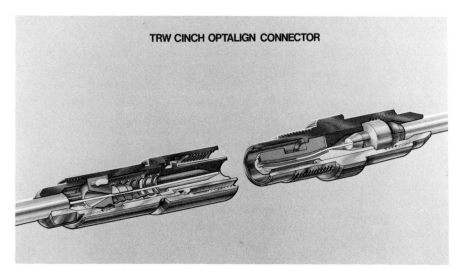

TRW CINCH OPTALIGN CONNECTOR

Figure 5.21   Optalign connector. (Courtesy of TRW Cinch.)

ray. Basically, four glass rods are placed about a common centerline and fused together to form a diamond-shaped hollow (four V-grooves). The four-rod array is then slightly curved on either end, leaving a straight section in the middle third of its length. Either end is conically flared to guide an entering fiber from either end. In that either end is slightly curved, an entering fiber is forced to glide along a common V-groove and arrive in the straight section a bit off axis. An entering fiber from the opposite end also arrives in the same V-groove in the center section. Since the straight section and its common V-groove now contain both fibers in longitudinal and angular alignment one need only butt the ends to achieve a low-loss, dry connection.

Unlike precision-machined tubes and bushings, the Optalign does not require precision of its envelope. One-half of the connector pair carries the four-rod glass guide in an appropriate plastic slug. The other half houses a spring-loaded piston which protects a free-standing fiber.

The intent is to prepare and strip a fiber to length so that it resides about halfway into the four-rod guide. The other half, containing the free-standing fiber, when coupled, will allow the piston to be pushed back, forcing the fiber into the glass guide and eventual alignment.

Two symmetrical pieces used in either connector half form a scissor clamp which secures the cable jacket to the connector shell, the cable strength members to the connector shell, and the fiber to prevent movement within the connector.

Resilient ferrules [1]. AMP's connector is a resilient alignment mechanism that absorbs variations in both the connector and the fiber. The single-position connector terminates a variety of fibers and can be used for fiber-to-fiber, fiber-to-source, and fiber-to-detector connections. The connector contains the following parts:

1. *Ferrule:* holds the fiber and serves as the resilient part of the alignment mechanism.
2. *Cap:* provides the means of maintaining the connection.
3. *Crimp ring:* provides the pressure to hold the fiber in the ferrule.
4. *Polishing bushing:* used to polish the fiber to the correct length and finish; it is discarded after use.

The connector is mated either to a splice bushing for fiber-to-fiber connections or to an input/output (I/O) bushing containing an active device. It may also be mated to the active-device connector described later.

The ferrule is the resilient alignment mechanism. The ferrule and bushing achieve the alignment. The inside of the bushing is tapered, and as the cap is screwed on, the taper compresses the ferrule. This compression moves the fiber on-center. When both connectors are joined in the bushing, the fibers are aligned, as shown in Fig. 5.22. The compression of the ferrule seals

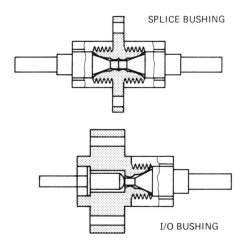

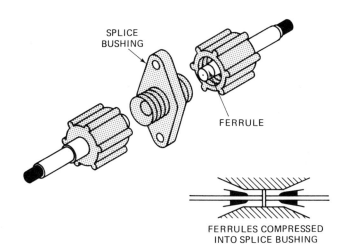

Figure 5.22   Resilient ferrule alignment mechanism. (From Ref. 1; courtesy of AMP.)

the fiber-to-fiber interface by providing a tight fit between the ferrule and fiber and between the ferrule and bushing. The hole in the ferrule is typically slightly larger than the fiber, since the ferrule will be compressed tightly around the fiber. Differences in fiber sizes are thus compensated for.

Ferrules are available for a variety of fiber sizes. They are color-coded according to their hole size. Except for the ferrule, the parts of the single-position connector are common to all. A typical termination, shown in Fig. 5.23, begins by stripping a half-inch of the fiber's jacket. If required (by the fiber manufacturer), epoxy is applied to the fiber. The fiber is slid into the

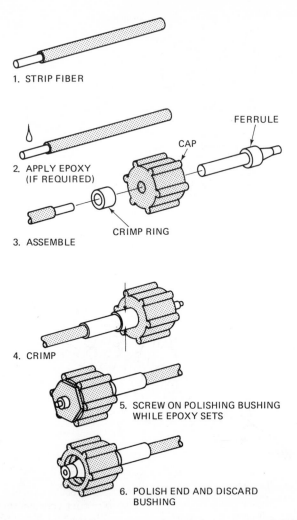

1. STRIP FIBER

FERRULE

CAP

2. APPLY EPOXY (IF REQUIRED)

CRIMP RING
3. ASSEMBLE

4. CRIMP

5. SCREW ON POLISHING BUSHING WHILE EPOXY SETS

6. POLISH END AND DISCARD BUSHING

Figure 5.23 Termination sequence. (From Ref. 1; courtesy of AMP.)

assembled connector and protrudes from the front of the ferrule. The crimp ring is crimped to the ferrule. The ring has two purposes: providing mechanical pressure to hold the fiber in the ferrule, and retaining the cap. After crimping, the polishing bushing is screwed onto the cap. The epoxy is allowed to set, and the fiber and ferrule are polished flush with the bushing. After polishing, the bushing is removed and discarded. The connector is ready for use.

Figure 5.24 shows typical bushings. Splice bushings are either free-hanging or bulkhead-mounted. I/O bushings, although generally similar, are available in a variety of styles to fit different needs. Some have different cavity dimensions to accommodate different TO cans; one style fits a molded-

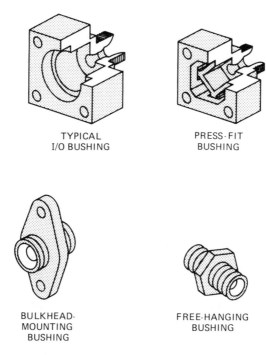

TYPICAL
I/O BUSHING

PRESS-FIT
BUSHING

BULKHEAD-
MOUNTING
BUSHING

FREE-HANGING
BUSHING

Figure 5.24   Bushings. (From Ref. 1; courtesy of AMP.)

lens package. Another style uses a press fit to hold the active device firmly. The I/O bushings are mounted over the active device and secured to the printed-circuit board.

AMP's small-fiber connector terminates 125 to 245-μm fibers. Like the single-position connector, the small-fiber connector uses a resilient ferrule to achieve alignment. To ensure the alignment of such small fibers, AMP has developed a proprietary method of placing the hole in the ferrule so that the fiber is centered when it is engaged.

The connector consists of the following parts:

1. *Cap:* provides the means of maintaining the connection.
2. *Retaining sleeve:* provides the pressure to hold the fiber in the ferrule; comes attached to the cap.
3. *Ferrule:* holds the fiber and serves as the resilient part of the alignment mechanism.
4. *Heat-shrink tubing:* shrunk around the retaining sleeve and fiber jacket to give extra support and strain relief.
5. *Polishing bushing:* used to polish the fiber to the correct length and finish.

In a typical termination (Fig. 5.25) the fiber's strength members are placed over the rear end of the ferrule. The retaining assembly (cap and sleeve) is slid over the strength member, and the sleeve is crimped to the strength members. The shrink tubing is then applied. The tubing gives additional reinforcement for holding the fiber in place and preventing sharp bending at the rear of the sleeve.

The hex-shaped cap shown in Fig. 5.25 is metal to provide shielding of the source and detector. Both the ferrule and the metal retaining sleeve vary according to the fiber used. Ferrules are color coded.

Although the small-fiber connector is intended to be used with the active-device connector, it is also compatible with the bushings used with the single-position connector.

The Optimate active-device connector, shown in Fig. 5.26, is an improved design for connecting a ferrule active device to the small-fiber or single-position connector. The active device is held in the connector by a press-on retention plate. The inside of the connector is tapered at both ends. Since the semiconductor is connected to the end of the ferrule by a fiber, the final alignment is an efficient one of fiber to fiber.

An AMP-FIT hand tool with the CERTI-CRIMP ratchet is used to crimp Optimate ferrules, except for small-fiber ferrules. The tool has two crimping areas: one for single-position and multiple-position ferrules and one for Optimate Multimate ferrules. A CHA-MP hand tool is also available for all these ferrules. The small-fiber connector is crimped with an AMP commercial hand tool with removable die sets. Figure 5.27 shows the three hand tools.

Note that most single-fiber connectors (except for TRW Cinch, for one) require the fibers to be epoxied in position. These so-called "wet connections" are much easier to make in the lab than in the field. After the epoxy has set, the fiber end (except for TRW Cinch, for one) must be polished square and smooth. In some cases a 50× microscope is necessary for checking the quality of the polish. The fiber end must be flush with the connector butt. For minimum losses the fiber ends should touch when the connectors are joined; however, in practice a gap must be maintained between the two surfaces, as repeated matings and vibrations will chip and scratch the surfaces. Any such damage to the surfaces would cause coupling loss. A typical gap distance is 0.0005 in.

5.3 COUPLERS

In the systems we have considered so far, only two terminals are used: a transmitting terminal and a receiving terminal. For multiterminal systems, a single common transmission path or line—a data bus—can be used instead of running a separate line from each terminal to each of the other terminals. Thus, the amount of cable required is significantly reduced.

Multiple signal transmission on a common bus is made possible through

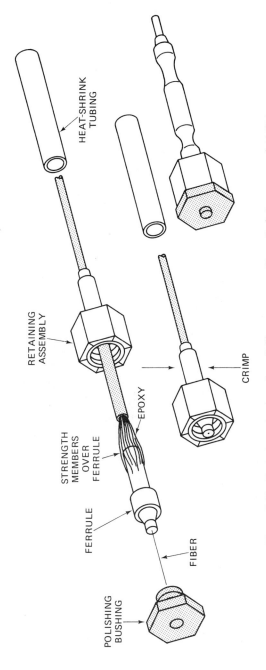

Figure 5.25 Typical termination of small-fiber connector. (From Ref. 1; courtesy of AMP.)

HEAT-SHRINK TUBING

RETAINING ASSEMBLY

EPOXY

STRENGTH MEMBERS OVER FERRULE

FERRULE

FIBER

POLISHING BUSHING

CRIMP

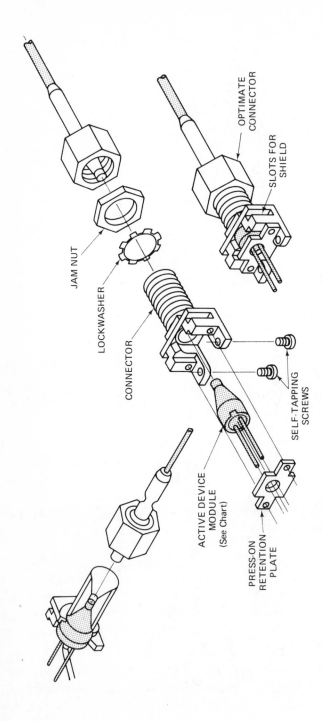

JAM NUT

LOCKWASHER

CONNECTOR

ACTIVE DEVICE
MODULE
(See Chart)

PRESS-ON
RETENTION
PLATE

SELF-TAPPING
SCREWS

OPTIMATE
CONNECTOR

SLOTS FOR
SHIELD

ACTIVE DEVICES AVAILABLE FROM MOROROLA SEMICONDUCTOR PRODUCTS

DESIGNATION	DESCRIPTION	RESPONSIVITY	RISE TIME	POWER OUTPUT
MFOE 102F	LED	–	25 ns	70 μW at 50 mA
MFOD 102F	PIN Photodiode	2 μA/μW	1 ns	–
MFOD 202F	Phototransistor	100 μA/μW	2.5 μs	–
MFOD 302F	Photodarlington	2.8 mA/μW	40 μs	–
MFOD 402F	Integrated Detector-Preamplifier	1 mV/μW	20 ns	–

Figure 5.26 Active-device connector. (From Ref. 1; courtesy of AMP.)

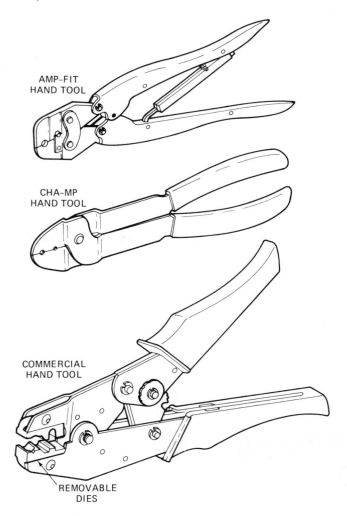

AMP-FIT
HAND TOOL

CHA-MP
HAND TOOL

COMMERCIAL
HAND TOOL

REMOVABLE
DIES

Figure 5.27   Crimping tools for connectors. (From Ref. 1; courtesy of AMP.)

multiplexing of the signals. Access to the bus is made through couplers which divert part of the signal power on the bus to the terminals and couple signals from the terminals onto the bus.

An optical coupler mixes optical signals or splits them as needed. Two basic types of couplers are star couplers and directional couplers.

5.3.1 Star Couplers*

Star couplers mix input light signals and then divide them equally among 2 to 19 output ports. Therefore, they may be used to combine numerous

---

* Most of Sections 5.3.1 through 5.3.4 are taken from Ref. 13.

MIXING REGION

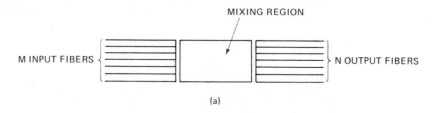

M INPUT FIBERS   N OUTPUT FIBERS

(a)

MIXING REGION   REFLECTING SURFACE

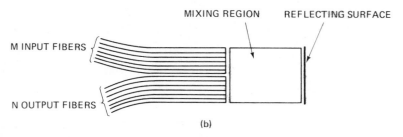

M INPUT FIBERS

N OUTPUT FIBERS

(b)

Figure 5.28   Star couplers: (a) transmissive star coupler; (b) reflective star coupler. (From Ref. 13; courtesy of ITT.)

signals, split one signal into numerous parts, or to insert optical power into or out of a fiber-optic link.

As shown in Fig. 5.28, there are two basic types of star couplers: the transmissive star and the reflective star. Either type of star has a set of input

Figure 5.29 Nine-channel distributive star coupler. (Courtesy of ITT Cannon.)

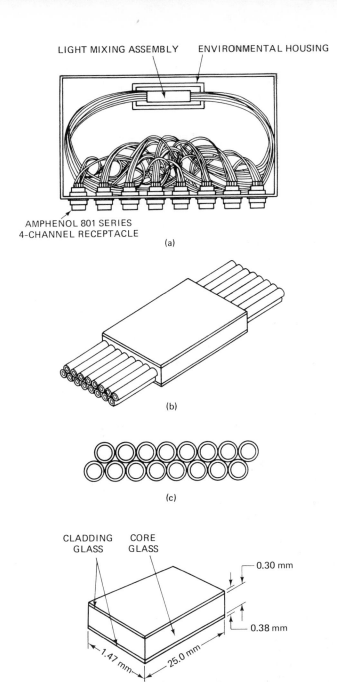

LIGHT MIXING ASSEMBLY   ENVIRONMENTAL HOUSING

AMPHENOL 801 SERIES
4-CHANNEL RECEPTACLE
(a)

(b)

(c)

CLADDING   CORE
GLASS   GLASS

0.30 mm

0.38 mm

1.47 mm

25.0 mm

(d)

Figure 5.30 Sixteen-channel transmissive star coupler: (a) 16-channel single-fiber transmissive star coupler; (b) optical fiber—rectangular light-guide assembly; (c) arrangement of the optical fibers for assembly to the rectangular light guide; (d) rectangular light guide. (From Ref. 14; courtesy of Amphenol, Division of Bunker Ramo.)

TABLE 5.4   ITT Directional Couplers

DIRECTIONAL COUPLER TYPE	PORT DIAGRAM	ADVANTAGES
Optical Directional Optical TDR Optical Monitor	1 ——→ 2 3	Totally optical tapoff
Optoelectronic	1 ——→ 2 PIN Photodiode 3	Allows direct electrical tapoff
Wavelength Duplex	1 ——← 2 0.85 μm 3 1.06 μm	Permits duplex transmission on single fiber

Source: Ref. 13; courtesy of ITT.

fibers, a set of output fibers, and a mixing region. The primary difference between the two types is that the transmissive star's input ports are optically isolated from each other, whereas the reflective star ports are all coupled. In the reflective star, any port can be used as input or output.

An advantage of the reflective star is that the relative number of input and output ports may be selected or varied after the device has been constructed. In the transmissive star, however, the number of input and output fibers is fixed by initial design and fabrication. On the other hand, the reflective star is usually less efficient, since a portion of the light fed into the coupler is injected back into the input fibers. With the same number of input and output ports, the transmissive star is twice as efficient as the reflective star.

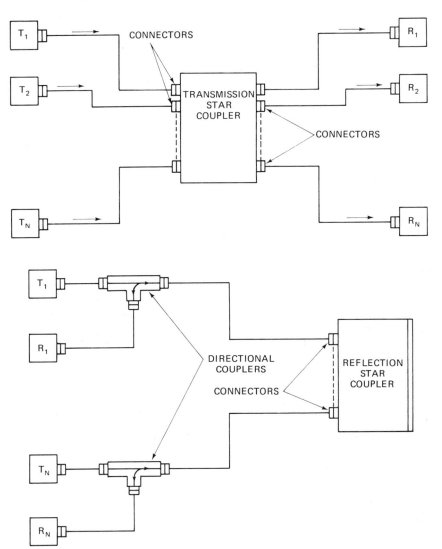

Figure 5.31 Star bus configurations. *T*, source; *R*, receiver. (From Ref. 13; courtesy of ITT.)

Star couplers (Fig. 5.29) have been built with cylindrically shaped transparent rods, fibers joined with a clear optical adhesive, rectangular waveguides, and focusing components.

Amphenol's 16-channel 32-port reflective star coupler uses a rectangular waveguide as the mixing device, as shown in Fig. 5.30 [14]. An advantage of this configuration is said to be the uniformity of the light at the output end. Test results were:

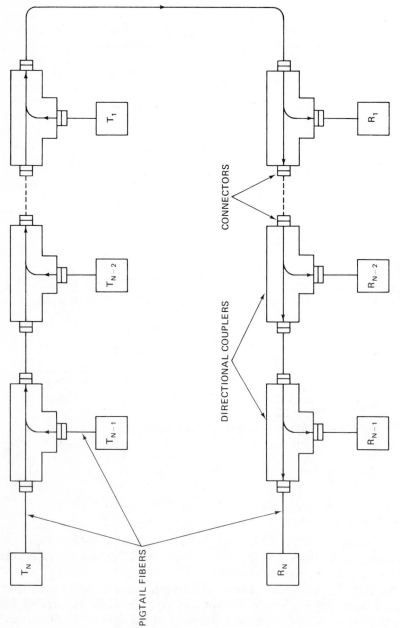

Figure 5.32 Loop bus configuration. (From Ref. 13; courtesy of ITT.)

CONNECTORS

DIRECTIONAL COUPLERS

PIGTAIL FIBERS

average port-to-port loss = 17.6 dB

average insertion loss   =   5.6 dB

optical signal range   =   3.3 dB

directivity   = approx. 9.8 dB

5.3.2 Directional Coupler

Directional couplers can be used as drop/insert, monitoring, branching, wavelength-duplexing, or mixing elements in multiple-access, bidirectional point-to-point, or data bus systems.

Three types of directional couplers developed by ITT are shown in Table 5.4. Two tapoff ratios are provided: a −30-dB tapoff for monitor applications, and a 50–50 power split for data bus and other special applications.

5.3.3 Data Bus Configuration

The basic data bus configurations are the star (Fig. 5.31) and the loop (Fig. 5.32). In loop configurations all information passes through each node; drop and insert elements are required to access subscriber terminals. In star configurations, a centralized selection process that is implemented by a branching element is required.

Both transmissive and reflective star configurations are shown in Fig. 5.31. The reflective star configuration requires only half the cable of the transmissive star configurations; however, 9 dB of power margin is sacrificed and an additional directional coupler is required for each channel. An alternative reflective star configuration is a single port for each transmitter and receiver.

Advantages of the loop configuration include expandability, ease of synchronization, and minimum cable cost. However, the loop configuration is severely limited by the accumulated insertion loss as a data stream passes successive drop nodes.

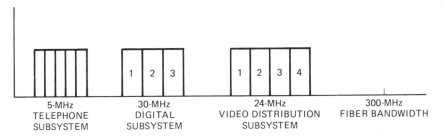

5-MHz	30-MHz	24-MHz	300-MHz
TELEPHONE	DIGITAL	VIDEO DISTRIBUTION	FIBER BANDWIDTH
SUBSYSTEM	SUBSYSTEM	SUBSYSTEM	

Figure 5.33   Frequency-domain allocations. (From Ref. 13; courtesy of ITT.)

5.3.4 Data Bus Multiplexing

Multiplexing techniques used in conventional wideband multiuser radio-frequency networks can also be employed in fiber-optic systems. Two common methods are time-division multiplexing (TDM) and frequency-division multiplexing (FDM).

Figure 5.33 shows an example of how individual services can be segregated by frequency-division multiplexing. The extremely high fiber bandwidth enables many channels to be multiplexed on a single, high-capacity fiber bus.

REFERENCES

1. "Introduction to Fiber Optics and AMP Fiber-Optic Products," AMP HB 5444, AMP Incorporated, n.d.

2. Stephan Ohr, "Splice System Cuts Field Fiber Losses to 0.15 dB," *Electronic Design,* Vol. 28, No. 10, May 10, 1980, p. 44

3. William E. Forbrich, "Aerial, Buried and Duct Fiber Optic Cable Go into the Field," *Telephony,* Dec. 25, 1978.

4. Robert Gallawa, "Optical Systems: A Review. 1. U.S. and Canada: Initial Results Reported," *IEEE Spectrum,* Oct. 1979, pp. 71–74.

5. "Glass-on-Glass Optical Fiber End Preparation," ITT Technical Note R–4, Dec. 78.

6. Thomas G. Giallorenzi, "Optical Communications Research and Technology," *Proceedings of the IEEE,* Vol. 66, No. 7, July 1978, pp. 744–780.

7. Detlef Gloge and others, "Fiber Splicing," in *Optical Fiber Telecommunications,* ed. Stewart E. Miller and Alan G. Chynoweth (New York: Academic Press, Inc., 1979), p. 465.

8. "Strengthening of Fiber Splices and Cable Described at Wire and Cable Symposium," *Laser Focus,* Feb. 1980, p. 84.

9. John N. Kessler, "Fiber-Optic Connectors: Prices Drop, Performance Rises," *Electro-Optical Systems Design,* Oct. 1979, p. 32.

10. L. M. Borsuk, "Fiber Optic Interconnections—A Tutorial Overview," ITT Cannon Electric, Santa Ana, Calif.

11. Ron Schultz, "Optalign Fiber Optic Connector," TRW Cinch, Aug. 1979.

12. W. S. Hudspeth, "Fiber Optic Connectors—Still a Budding Technology," *Electro-Optical Systems Design,* Oct. 1978, p. 48.

13. "Applications of ITT Optical Fiber Multiple Access Couplers," ITT Technical Note R-8, Aug. 1978.

14. Lawrence J. Coyne, "Distributive Fiber Optic Couplers Using Rectangular Lightguides as Mixing Elements," Bunker Ramo Amphenol, Danbury, Conn. Presented at the Second International Fiber Optics and Communications Exposition, Sept. 5–7, 1979; courtesy of Information Gatekeepers, Inc., Brookline, Mass.

6

RECEIVERS

At the receiver end of a fiber-optic system, light waves must be converted to electrical output signals and then amplified for further use. In the case of digital modulation, these signals must then be processed to recover the original information.

Semiconductor light sensors (photodetectors) are used to convert the optical energy to electrical energy. The detectors most commonly used in fiber optics are silicon positive-intrinsic-negative (PIN or p-i-n) photodiodes, avalanche photodiodes (APD), phototransistors, and photodarlington transistors.

Alternative detectors, such as thyristor-type *pnpn* devices and photosensitive GaAs MOSFETs, are under development but can be ignored for now.

Although phototransistors and photodarlington transistors have a higher gain than photodiodes, they have a low bandwidth (150 kHz). Therefore, although material will be presented on these transistors, we concentrate on PINs and APDs in this chapter.

In general, APDs cost 5 to 10 times more than PINs, but in exchange they offer higher sensitivity (10 to 20 dB). In addition, APDs have a greater bandwidth (3 GHz) [1] than PINs. A disadvantage is that APDs require high voltage (400 volts) compared to PINs (40 volts) and have greater noise. Also, special feedback circuits are necessary for APDs to reduce their sensitivity to temperature changes.

For these reasons, APDs are most often used in long-distance links and where receiver sensitivity at a high modulation frequency is important. PINs are useful in those short links where cost is an important consideration.

For the 800- to 900-nm region, silicon devices are preferred. However, for the 1100- to 1700-nm region, the region of the future, photodetectors are still under development and no one material has won out. Silicon's efficiency is too poor at these wavelengths. Germanium is the best bet for now, but indium-gallium-arsenide-phosphide may prove superior.

For use in fiber optics, photodetectors must be highly sensitive (responsive) to weak light signals yet have sufficient bandwidth or speed of response to handle the incoming signals. At the same time the detectors must be relatively immune to changes in temperature and must add a minimum of noise to the circuit.

The response of a photodetector is expressed by the term *responsivity* (*R*), which is the ratio of output current to the incident optical flux. It is measured in amperes per watt (A/W).

One of the main factors that determine sensitivity is the noise in the circuit. In any fiber-optic system there is a certain amount of spurious, unwanted energy called *noise*. It can be caused by electromagnetic interference, crosstalk, and other phenomena. In light-related devices, there is always a small trickle of current, called the *dark current,* which flows when there is no light in the circuit. Thus, any circuit has a prevailing noise level, a certain level

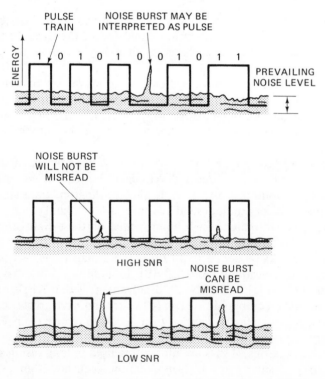

Figure 6.1   Electrical noise. (From Ref. 2; courtesy of AMP.)

of unwanted electrical energy. Additional bursts of energy will cause noise above the prevailing level [2] as shown in Fig. 6.1.

If the signal's strength is less than the prevailing noise level, the signal will be lost in the noise. At the same time, extra bursts of noise may be interpreted as part of the signal, as a pulse, for example [2].

Two terms are used to describe the relation between noise and the signal: *signal-to-noise ratio* (SNR or S/N) and *bit-error-rate* (BER). In both analog and digital systems SNR expresses how much stronger the signal is than the noise. For example, in a digital system, an SNR of 14.3 dB means that the signal is 26.9 times larger than the noise level. This SNR corresponds to 1 error for every 1 million bits transmitted (for certain circuit designs). For a rate of 1 error in 1 billion, the SNR must be 16.7 dB. Such errors can be expressed more straightforwardly with BER, which is the ratio of incorrect bits to the total bits received. A BER of 10^{-6} means 1 error per 1 million bits. One error per 1 billion bits gives a BER of 10^{-9}. As bandwidth increases, so does the noise; higher bandwidths require a better SNR to maintain a constant BER [2].

In photodetector circuits, one way to reduce the problem of noise is to ensure that the incoming light signal is strong enough to create an acceptable current [6]. Still another way is to use error detection/correction logic, such as a simple check-sum routine.

6.1 PIN PHOTODIODES

In the photodiode family, there are two basic types: depletion-layer and avalanche. The two types are similar except that the avalanche type has a built-in gain mechanism.

Of the various depletion-layer types, only the PIN diode (Fig. 6.2) is significant for use in fiber optics. In fact, it is the most common light sensor in such circuits.

As shown in Figs. 6.3 or 6.4, the PIN photodiode has a layer of undoped or intrinsic (I) material sandwiched between a layer of positively (P) doped and negatively (N) doped material. Light enters the diode through a tiny "window" which is about the size of the fiber core.

In effect, the photodiode works just the opposite from an LED. Photons that fall on the carrier-void depletion region (Fig. 6.5) create carriers. Light-generated carriers in the depletion region allow current to flow through the diode. Notice that the photons are absorbed by the electrons in the valence band. This allows electrons to break free of the valence band and enter the conduction band. The result is a free electron and a hole, both carriers of current. When the light is removed, the depletion region is restored and current stops [2].

The PIN photodiode provides no gain or amplification. For every

Figure 6.2 RCA C-30920E PIN photo-diode. (Courtesy of RCA.)

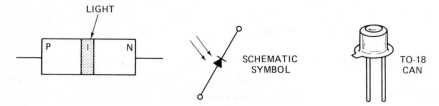

Figure 6.3 PIN photodiode. (From Ref. 2; courtesy of AMP.)

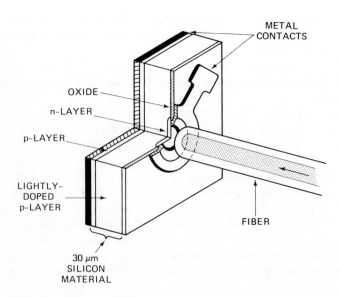

Figure 6.4 Cutaway view of PIN photodiode. (From Ref. 3; copyright 1975 Bell Laboratories RECORD.)

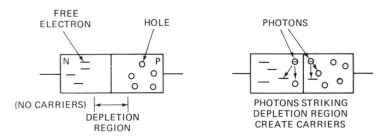

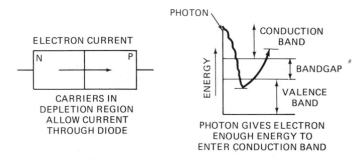

Figure 6.5 Operation of a photodiode. (From Ref. 2; courtesy of AMP.)

photon captured in the intrinsic layer, an electron–hole pair is set flowing as current. Since amplifiers can be added after the diode, the lack of amplification is not always a problem [2]. As shown in Fig. 6.6, the PIN requires a bias voltage, which may be as low as 5 V.

Tables 6.1 and 6.2 give the characteristics of a typical PIN (Fig. 6.7). Table 6.3 compares several commercially available PIN photodiodes.

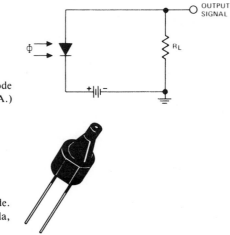

Figure 6.6 Basic solid-state photodiode circuit. (From Ref. 4; courtesy of RCA.)

Figure 6.7 MF0D104F PIN photodiode. (From Ref. 5; copyright 1979 Motorola, Inc.)

TABLE 6.1 Maximum Ratings for Motorola MFOD104F PIN Photodiode[a]

Rating	Symbol	Value	Unit
Reverse Voltage	V_R	100	V
Total device dissipation at $T_A = 25°$ C Derate above 25°C	P_D	100 0.57	mW mW/°C
Operating Temperature Range	T_A	-30 to $+85$	°C
Storage temperature range	T_{stg}	-30 to $+100$	°C

Source: Ref. 5; copyright 1979 Motorola Inc.
[a] $T_A = 25°C$ unless otherwise noted.

TABLE 6.2 Electrical and Optical Characteristics for Motorola MFOD104F PIN Photodiode[a]

Characteristic	Symbol	Min.	Typ.	Max.	Unit
Electrical Characteristics					
Dark current ($V_R = 20$ V, $R_L = 1.0$ M, $H \approx 0$)	I_D	—	—	2.0	nA
Reverse breakdown voltage ($I_R = 10$ μA)	BV_R	100	200	—	V
Forward voltage ($I_F = 50$ mA)	V_F	—	0.82	1.2	V
Total capacitance ($V_R = 5.0$ V, $f = 1.0$ MHz)	C_T	—	—	4.0	pF
Noise equivalent power	NEP	—	50	—	fW/√Hz
Optical Characteristics					
Responsivity at 900 nm ($V_R = 5.0$ V, P $= 10$ μW[b])	R	0.15	0.40	—	μA/μW
Response time at 900 nm $V_R = 5.0$ V 12 V 20 V	t_{on}, t_{off}	— — —	6.0 4.0 2.0	— — —	ns
Numerical aperture of input port, 3.0 dB [200-μm (8-mil)-diameter core]	NA	—	0.48	—	—

Source: Ref. 5; copyright 1979 Motorola Inc.
[a] $T_A = 25°C$
[b] Power launched into optical input port. The designer must account for interface coupling losses.

TABLE 6.3   Commercially Available PIN Photodiodes

Device	Responsivity (A/W) at λ_0 (nm)	3-dB Bandwidth (MHz)	Dark Current (nA at V)	Active Area (mm²)	Package
RCA C30807	0.6 at 900 0.15 at 1060	60	2 at 10 10 at 45	0.8	TO-18
RCA C30808	0.6 at 900 0.15 at 1060	45	5 at 10 30 at 45	5.0	TO-5
RCA C30809	0.6 at 900 0.15 at 1060	25	25 at 10 70 at 45	50.0	TO-8
RCA C30810	0.6 at 900 0.15 at 1060	20	80 at 10 300 at 45	100.0	RCA 25-mm
RCA C30822	0.6 at 900 0.15 at 1060	40	10 at 10 50 at 45	20.0	TO-8
RCA C30831	0.6 at 900 0.15 at 1060	60	1 at 10 10 at 45	0.2	TO-18
RCA C30812	0.60 at 900 0.5 at 1060	25	20 at 20 30 at 200	5.0	TO-5
UDT PIN-020A	0.42 at 850	75	0.5 at 10	0.2	TO-18
EG & G SGD-040	0.5 at 900	120	5 at 100	12.6	TO-46
Thomson CSF TCO-202	0.4 at 830		2 at 10	0.5	Special// F/O
HP 5082-4205	0.5 at 770	360	0.15 at 10	0.02	Pill
BNR D-5-2	0.55 at 840	120	1 at 45	0.012	TO-1811// F/O
RCA C309OGE	0.6 at 900	60	10 at 10 100 at 90	5.0	TO-5
NEC LSD 39A	0.3 at 633	1800	0.2		
NEC LSD 39B	0.3 at 633	900	1.0		

Source: Reprinted with permission from Ref. 6, p. 167 (*Electronic Design,* Vol. 28, No. 9); copyright Hayden Publishing Co., Inc., 1980.

6.2 AVALANCHE PHOTODIODES

In contrast with the PIN photodiode, the avalanche photodiode (APD) (Fig. 6.8) has internal gain. This gain is a result of impact ionization, which occurs at high reverse-bias voltages, near the breakdown voltages. During impact ionization a free electron or hole can gain sufficient energy to ionize a bound electron. The ionized carriers cause further ionizations, leading to an avalanche of carriers [6, p. 167].

As shown in Fig. 6.9, the standard APD is a "reach-through" diode.

Figure 6.8 RCA C30904E avalanche photodiode. (Courtesy of RCA.)

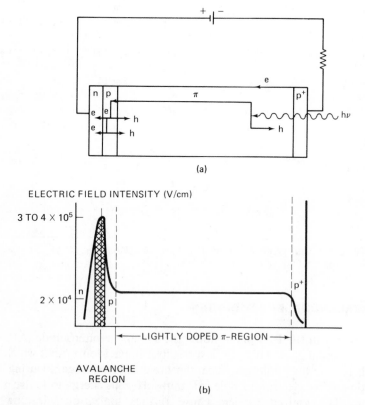

Figure 6.9 Operation of an avalanche photodiode. [Reprinted with permission from Ref. 6, p. 16 (*Electronic Design,* Vol. 28, No. 9); copyright Hayden Publishing Co., Inc., 1980.]

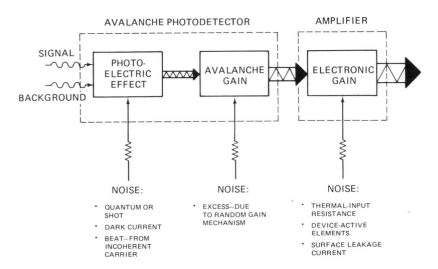

The depletion region has a wide drift region and a narrow multiplying region, as indicated in the electric field profile. Photons are absorbed in the lightly doped π-region, whereas photogenerated carriers cause impact ionization in the avalanche region [6].

When the peak electric field is 5 to 10% below the avalanche breakdown field, the doping levels of the p and n avalanche regions allow the diode depletion layer to "reach through" to the low-doped π-region [6].

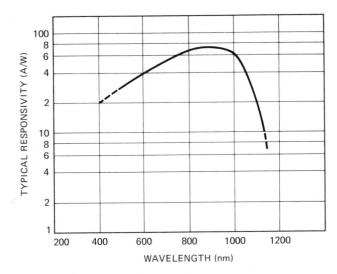

Figure 6.11   Typical spectral responsivity characteristics for RCA C309trE avalanche photodiode. (From Ref. 8; courtesy of RCA.)

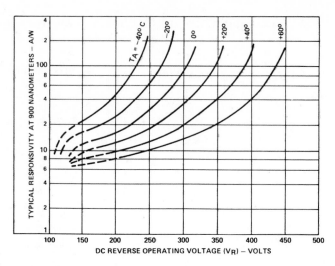

Figure 6.12 Typical responsivity at 900 nm vs. operating voltage for RCA C30954E avalanche photodiode. (From Ref. 8; courtesy of RCA.)

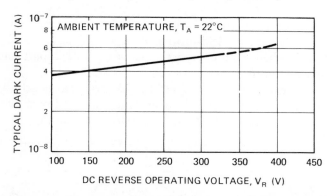

Figure 6.13 Typical dark current vs. operating voltage for RCA C30954E avalanche photodiode. (From Ref. 8; courtesy of RCA.)

The various noise sources for an APD are shown in Fig. 6.10. Parameters for a typical developmental type APD are given in Figs. 6.11 through 6.13 and Tables 6.4 and 6.5.

6.3 PHOTOTRANSISTORS

In effect, a phototransistor is a junction transistor that is exposed to illumination. Light applied to the base (Fig. 6.14) creates carriers in the base and allows the transistor to turn on. Note that the phototransistor not only converts light to electricity, but amplifies it as well. Because of this amplification

**TABLE 6.4   Maximum Rating, Absolute-Maximum Values
RCA C30954E Avalanche Photodiode**

Reverse bias current	200 max.	μA
Photocurrent density j_p, at 22°C		
Average value, continuous operation	5	mA/mm}
Peak value	20	mA/mm^2
Forward current, I_F, at 22°C		
Average value, continuous operation	5 max.	mA
Peak value (for 1 s	50 max.	mA
duration, nonrepetitive)		
Maximum total power dissipation at	0.1 max.	W
22°C (with heat sink cooling		
provided to case)		
Ambient temperature		
Storage, T_{stg}	− 60 to + 100	°C
Operating T_A	− 40 to + 70	°C
Soldering		
for 5 s	200	°C
	(Leads only)	

Source: Ref. 8; courtesy of RCA.

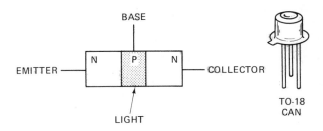

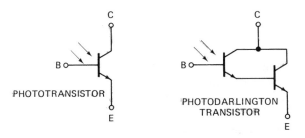

Figure 6.14   Phototransistors. (From Ref. 2; courtesy of AMP.)

141

TABLE 6.5 Electrical Characteristics of RCA C30954E Avalanche Photodiode

Electrical Characteristics at T_A = 22°C At the dc Reverse Operating Voltage V_R Supplied with the Device				
		C30954E Light Spot Diameter 0.25 mm (0.01 in.)		
	Min.	Typ.	Max.	Units
Breakdown voltage, V_{BR}	300	375	475	V
Temperature coefficient of V_R for constant gain	—	2.2	—	V/°C
Gain	—	120	—	
Responsivity				
At 900 nm	65	75	—	A/W
At 1060 nm	30	36	—	A/W
At 1150 nm	4	5	—	A/W
Quantum efficiency				
At 900 nm	—	85	—	%
At 1060 nm	—	36	—	%
At 1150 nm	—	5	—	%
Total dark current, I_d	—	5×10^{-8}	1×10^{-7}	A
Noise current i_n f = 10 kHz, $\triangle f$ = 1.0 Hz	—	1×10^{-12}	2×10^{-12}	A/Hz$^{1/2}$
Capacitance, C_d	—	2	4	pF
Series resistance	—	—	15	Ω
Rise time, t_r R_L = 50 Ω, λ = 900 nm, 10% to 90% points	—	2	3	ns
Fall time R_L = 50 Ω, λ = 900 nm, 90% to 10% points	—	2	3	ns

Source: Ref. 8; courtesy of RCA.

it has more gain than a photodiode; however, it has a more limited bandwidth.

The bias circuit for the phototransistor is more complex than that for a PIN photodiode but less complex than that for an APD. In the photodarlington transistor, two transistors are connected on the same chip to provide even more gain.

Specifications for an NPN silicon phototransistor are given in Tables 6.6 and 6.7. An NPN silicon photodarlington transistor is described in Tables 6.8 and 6.9.

TABLE 6.6   Maximum Ratings for Motorola MFOD202F Phototransistor[a]

Rating	Symbol	Value	Unit
Collector–emitter voltage	V_{CEO}	50	V
Emitter–base voltage	V_{EBO}	10	V
Collector–base voltage	V_{CBO}	50	V
Light current	I_L	250	mA
Total device dissipation at $T_A = 25°C$	P_D	250	mW
Derate above 25°C		1.43	mW/°C
Operating temperature range	T_A	− 30 to + 85	°C
Storage temperature range	T_{stg}	− 30 to + 100	°C

Source: Ref. 9; copyright 1979 Motorola, Inc.
[a] $T_A = 25°C$ unless otherwise noted.

TABLE 6.7   Characteristics of Motorola MFOD202F Phototransistor[a]

Characteristic	Symbol	Min.	Typ.	Max.	Unit
Static Electrical Characteristics					
Collector dark current	I_{CEO}	—	5.0	50	nA
($V_{cc} = 2V$, $H \approx 0$)					
Collector–base breakdown voltage	BV_{CBO}	50	—	—	V
($I_C = 100 \mu A$)					
Collector–emitter breakdown voltage	BV_{CEO}	50	—	—	V
($I_C = 100\mu A$)					
Optical Characteristics					
Responsivity	R	70	100	—	$\mu A/\mu W$
($V_{CC} = 20$ V, $R_L = 10 \, \Omega$, $\lambda \approx 900$ nm, $P = 1.0 \, \mu W$[b])					
Photo current rise time	t_r	—	2.5	—	μs
($R_L = 100 \, \Omega$)					
Photo current fall time	t_f	—	4.0	—	μs
($R_L = 100 \, \Omega$)					
Numerical aperture of input port	NA	—	0.48	—	—
[200-μm (8-mil)-diameter core]					

Source: Ref. 9; copyright 1979 Motorola, Inc.
[a] $T_A = 25°C$ unless otherwise noted.
[b] Power launched into optical input port. The designer must account for interface coupling losses.

6.4 RECEIVERS

The output current of the photodetector is likely to be quite feeble. For a light output of 10 nW, for example, the detector output current may be as low as 5 nA for a PIN photodiode and 0.5 μA for an APD [11, p. 633].

It is the job of the receiver to amplify this current and recover either the

TABLE 6.8 Maximum Ratings for Motorola MFOD300 Photodarlington Transistor[a]

Rating	Symbol	Value	Unit
Collector–emitter voltage	V_{CEO}	40	V
Emitter–base voltage	V_{EBO}	10	V
Collector–base voltage	V_{CBO}	50	V
Light current	I_L	250	mA
Total device dissipation at T_A = 25°C	P_D	250	mW
Derate above 25°C		1.43	mW/°C
Operating and storage junction temperature range	T_J, T_{stg}	−55 to +175	°C

Source: Ref. 10; copyright 1978 Motorola, Inc.
[a] T_A = 25°C unless otherwise noted.

TABLE 6.9 Characteristics of Motorola MFOD300 Photodarlington Transistor[a]

Characteristic	Symbol	Min.	Typ.	Max.	Unit
Static Electrical Characteristics					
Collector dark current (V_{CE} = 10 V, $H \approx 0$)	I_{CEO}	—	10	100	nA
Collector–base breakdown voltage (f_c = 100 μA)	BV_{CBO}	50	—	—	V
Collector–emitter breakdown voltage (I_C = 100 μA)	BV_{CEO}	40	—	—	V
Emitter–base breakdown voltage (I_E = 100 μA)	BV_{EBO}	10	—	—	V
Optical Characteristics					
Sensitivity (V_{CC} = 5.0 V, R_L = 10 Ω, $\lambda \approx$ 900 nm)	S_{rceo}	56	75	—	mA/mW· cm²
Photo current rise time[b] (R_L = 100 Ω)	t_r	—	40	—	μs
Photo current fall time[b] (R_L = 100 Ω)	t_f	—	60	—	μs

Source: Ref. 10; copyright 1978 Motorola, Inc.
[a] T_A = 25°C.
[b] For unsaturated response-time measurements, radiation is provided by pulsed GaAs (gallium–arsenide) light-emitting diode ($\lambda \approx$ 900 nm) with a pulse width equal to or greater than 500 μs, I_C = 1.0 mA peak.

digital or analog information from it. In the process it must not introduce large amounts of noise.

To reduce noise yet keep a large dynamic range, many fiber-optic receivers use a *trans-impedance* amplifier. This is a high-gain low-noise amplifier with a feedback circuit and a low input impedance [3, p. 339; 12].

The basic block diagrams for digital and analog receivers are shown in Figs. 6.15 and 6.16, respectively.

In a digital receiver, the regenerator compares the filter's output voltage to a threshold once per time slot (pulse interval) to determine if a pulse is present. The regenerator is synchronized to the rate of arrival of the digital pulses [11, p. 629].

Figure 6.17 shows a variation of the basic digital receiver. As you can see, a fiber-optic receiver is more complex than a fiber-optic transmitter. Integrated detector–preamplifiers are shown in Figs. 6.18 and 6.19.

The Motorola MFOD403F integrated detector/preamplifier (Figs. 6.20 and 6.21) is designed for medium-bandwidth (up to 50 Mbaud) medium-distance systems. It requires a minimum of 2.0 μW of light for a 10-dB SNR.

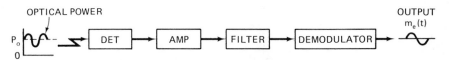

Figure 6.15   Digital receiver black diagram. (From Ref. 11, p. 629; copyright 1979 Bell Telephone Laboratories, Inc.; reprinted by permission.)

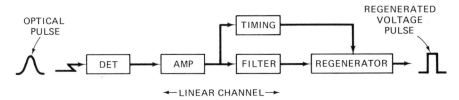

Figure 6.16   Analog receiver block diagram. (From Ref. 11, p. 630; copyright 1979 Bell Telephone Laboratories, Inc.; reprinted by permission.)

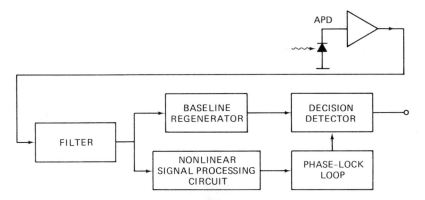

Figure 6.17   Digital receiver block diagram. (From P. Russer, "Introduction to Optical Communications," in *Optical Fibre Communications,* ed. M. J. Howes and D. V. Morgan; copyright 1980 John Wiley & Sons Ltd.; reprinted by permission of John Wiley & Sons Ltd.)

Figure 6.18 Integrated PIN photodiode and preamplifier. (Courtesy of RCA.)

Figure 6.19 Integrated avalanche photodiode and preamplifier. (Courtesy of RCA.)

Figure 6.20 Motorola MF0D403F integrated detector/preamplifier. (From Ref. 13; courtesy of Motorola, Inc.)

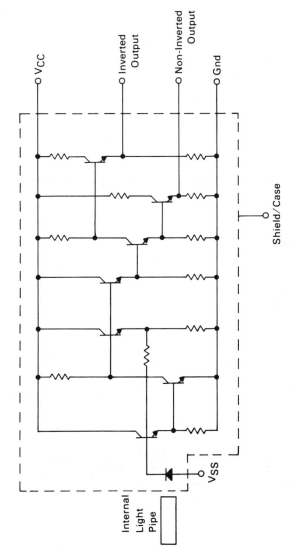

Figure 6.21 MFOD403F integrated detector/preamplifier equivalent schematic. (From Ref. 13; copyright 1979 Motorola, Inc.)

Although bias voltages from 0 to -20 V are usable, the more negative voltages yield faster maximum bit rates [13]. Maximum ratings for the unit are given in Table 6.10. Figure 6.22 shows its use in a complete receiver.

A simple design preamplifier is shown in Fig. 6.23. It includes only two active devices: a D MOSFET and a video amplifier. Dc coupling around the feedback loop stabilizes the circuit with respect to supply variations and spreads in active-device parameters [14].

With the aid of Fig. 6.24 we can examine a complete digital receiver on a block-by-block basis.*

The unit's optical detector behaves like a current source—whose magnitude depends on the level of incident optical energy—in parallel with a capacitor whose value is a function of device design and the magnitude of reverse bias. Because the detector is a high-impedance source with a tiny small-signal output, it is difficult to interface without introducing noise, RFI, and reactive loads, which degrade signal quality.

TABLE 6.10   MFOD403F Integrated Detector/Preamplifier Maximum Ratings[a]

Rating	Symbol	Value	Unit
Operating voltage	V_{CC}	7.5	V
Total device dissipation at $T_A = 25°C$[b]	P_D	250	m W
Derate above 25°C		1.43	m W / °C
Operating temperature range	T_A	-30 to $+85$	°C
Storage temperature range	T_{stg}	-30 to $+100$	°C

Source: Ref. 13; copyright 1979 Motorola, Inc.

[a] $T_A = 25°C$ unless otherwise noted.

[b] Package limitations.

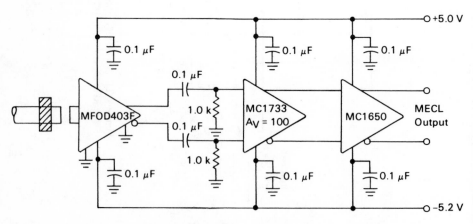

Figure 6.22   Ac-coupled 40-Mb/s MEDL output fiber-optic receiver. (From Ref. 13; copyright 1979 Motorola, Inc.)

* The following eight paragraphs are reprinted from Ref. 15; copyright 1980 Cahners Publishing Co., *EDN*.

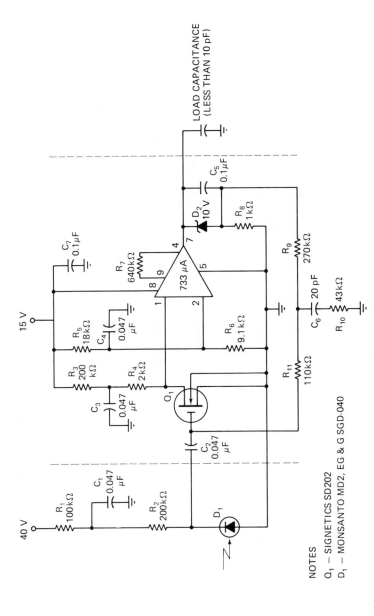

Figure 6.23 Simple design preamplifier. (From Ref. 14; copyright 1980 Cahners Publishing Co., *EDN.*)

NOTES

Q_1 – SIGNETICS SD202
D_1 – MONSANTO MD2, EG & G SGD-040

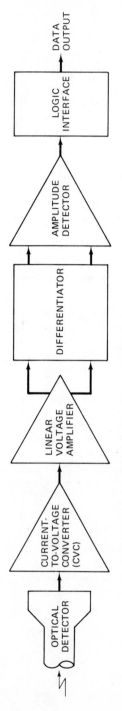

Figure 6.24 Digital receiver block diagram. (From Ref. 15; copyright 1980 Cahners Publishing Co., *EDN.*)

For this reason, the receiver's current-to-voltage converter (CVC) usually couples as closely as possible to the detector, with the interface between them often shielded from outside interference. Typically, a trans-impedance-amplifier circuit built from an op amp or other high-gain amplifier with negative current feedback, the CVC performs three functions:

1. It provides signal gain by producing an output voltage proportional to the input current.
2. It furnishes a low output impedance by virtue of its high open-loop gain and negative feedback.
3. It develops a virtual ground at its signal input.

Now that the incoming optical signal is a voltage pulse with fast rise and fall times produced by a low-impedance source, more conventional means can serve subsequent processing.

The linear voltage amplifier—the third element in the receiver's block diagram—starts this processing. It should have gain sufficient to amplify the CVC's expected noise nearly up to the minimum threshold level of the fifth block—the amplitude detector.

The voltage amplifier's bandwidth and rise time are also critical. The rise time of the signal appearing at the amplitude detector's input—which is

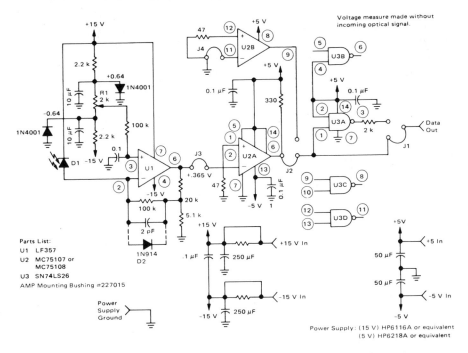

Figure 6.25   Megabit receiver. (From Ref. 16; courtesy of Motorola, Inc.)

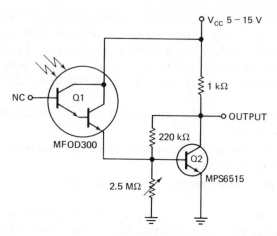

Figure 6.26 Low-frequency photodarlington receiver. (From Ref. 16; courtesy of Motorola, Inc.)

the voltage amplifier's output here—defines system rise time. The optical detector and preamp determine the bandwidth of a well-designed system.

The differentiator strips off duty-cycle-related baseline variations from the data stream. This eases the receiver's amplitude detector's decision tasks.

The amplitude detector must have two thresholds to handle the differentiator's output, which returns to the reference voltage level from either pulse polarity. A comparator or line receiver with positive feedback satisfies this requirement.

As the final block in the receiver, the logic interface serves to generate a

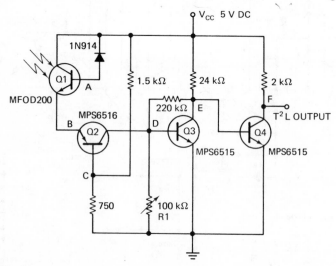

Figure 6.27 Twenty-kilobit phototransistor receiver. (From Ref. 16; courtesy of Motorola, Inc.)

standard logic level and provide sufficient drive capability to simplify interfacing. As noted, the amplitude detector actually generates the receiver's logic level. However, the interface block buffers the detector's output and provides some isolation from the outside world.

An experimental megabit fiber-optic receiver developed by Motorola is shown in Fig. 6.25.* The receiver may be connected for TTL or CMOS compatibility.

The receiver uses an MFOD100 PIN photodiode as an optical detector. The detector diode responds linearly to the optical input over several decades of dynamic range.

The PIN detector output current is converted to voltage by integrated circuit U1 (operational amplifier LF357). The minimum photocurrent required to drive U1 is 250 nA.

Receiver dynamic range is extended with diode D2 to prevent U1 from saturating at large optical power inputs.

Integrated circuit U2 acts as a voltage comparator. Its worst-case sensitivity of 50 mV determines the size of the pulse required out of U1. U2 detects, inverts, and provides standard TTL logic levels to the output.

CMOS compatible operation is available when integrated circuit U3 is wired into the printed-circuit board. This IC is an open-collector TTL quad, two-input NAND-gate device. Jumper wire J1 must be connected from U3 output to the receiver output terminal.

A simple photodarlington receiver (Fig. 6.26) may be used in a low-frequency system.

The output of the MFOD300 drives a single common-emitter amplifier (MPS6515). This circuit operates from a + 5- to + 15-V power supply, and its output is TTL- and CMOS-compatible.

The phototransistor receiver circuit shown in Fig. 6.27 may be used to data rates up to 20 kbits. The receiver sensitivity at 10 kbits is 4.7 μW.

REFERENCES

1. "Fiber Optics: Assessing New Technology," *Electronic Design,* Vol. 26, No. 22, Oct. 25, 1978, p. 58.

2. "Introduction to Fiber Optics and AMP Fiber-Optic Products," AMP HB 5444, AMP Incorporated, n.d.

3. Tingye Li, "Optical Transmission Research Moves Ahead," *Bell Laboratories RECORD,* Sept. 1975, pp. 333–340.

4. "Optical Communications Products," RCA OPT-115.

5. "MFOD104F Fiber Optics PIN Photo Diode," Motorola Advance Information, 1979.

6. Joseph Zucker, "Choose Detectors for Their Differences, to Suit Different Fiber-Optic Systems," *Electronic Design,* Vol. 28, No. 9., Apr. 26, 1980.

* The following eight paragraphs are from Ref. 16.

7. Stewart E. Miller, Tingye Li, and Enrique A. J. Marcatilli, "Research toward Optical-Fiber Transmission Systems; Part II: Devices and Systems Considerations," *Proceedings of the IEEE,* Vol. 61, No. 12, Dec. 1973, p. 1736.

8. "Photodiode Developmental Types," RCA Electro Optics and Devices, Oct. 1979.

9. "MFOD202F Fiber Optics NPN Silicon Phototransistor," Motorola Semiconductors Advance Information, 1979.

10. "MFOD300 Fiber Optics NPN Silicon Photodarlington Transistor," Motorola Semiconductors Advance Information, 1978.

11. Stewart D. Personick, "Receiver Design," in *Optical Fiber Telecommunications,* ed. Stewart E. Miller and Alan G. Chynoweth (New York: Academic Press, Inc., 1979).

12 Tien Pei Lee and Tingye Li, "Photodetectors," in *Optical Fiber Telecommunications,* ed. Stewart E. Miller and Alan Chynoweth (New York: Academic Press, Inc., 1979), p. 600.

13. "MFOD403F Fiber Optics Integrated Detector Preamplifier," Motorola Semiconductors Product Preview, 1979.

14. Yishay Netzer, "Simplifying Fiber-Optic Receivers with a High-Quality Preamp," *EDN,* Sept. 20, 1980, pp. 162, 163.

15. Vincent L. Mirtich, "Designer's Guide to: Fiber-Optic Data Links—Part 1," *EDN,* June 20, 1980, pp. 133–140.

16. "Basic Experimental Fiber Optic Systems," Motorola Semiconductors Advance Information, April 1978.

7

TYPICAL SYSTEMS

In this chapter we look at some off-the-shelf fiber-optic links and a fiber-optic system.

7.1 OFF-THE-SHELF FIBER-OPTIC LINKS

Complete ready-to-go links are available, off the shelf, for shorter links. With these links, it is not necessary to design or fabricate any circuits. Two representative types are described in the following paragraphs.

7.1.1 RCA Fiber-Optics Data Links*

RCA's fiber-optics data links consist of a transmitter and receiver, each in a small compact package. These systems (Fig. 7.1) offer the designer a choice of sensitivity and bandwidth, making them useful in a wide variety of applications, which include secure data transmission, high-voltage optically isolated data systems, computer interface, and process control data systems.

The transmitters contain an RCA high-speed GaAlAs LED and associated electronics for the drive circuit. A fiber-optic cable is internally coupled from the emitting region of the GaAlAs chip to an optical connector.

The receivers use an RCA silicon PIN photodetector with amplifier and threshold detection circuits to convert input light pulses to standard TTL electrical output signals.

* This section is taken from Ref. 1.

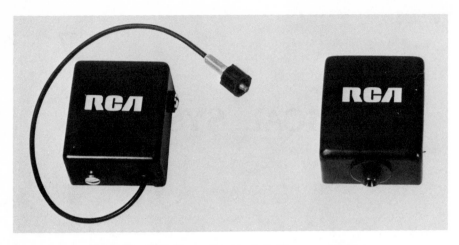

Figure 7.1   RCA fiber-optics data link. (Courtesy of RCA.)

They are capable of transmitting over optical fibers at data rates up to 20 Mb/s and over distances in excess of 1 km. As shown in Fig. 7.2, the RCA systems described herein employ pigtail IR edge-emitting diodes in the transmitter modules; the receivers contain a reverse-biased PIN photodiode with hybrid preamplifier for maximum sensitivity.

The IR emitter output may be varied (increased or decreased) slightly by adjusting the LED CURRENT ADJUST, accessible from the bottom of the transmitter module.

RCA can provide custom systems with preterminated fibers for operation over specific lengths. Specifications for the links are given in Tables 7.1 and 7.2.

7.1.2 Math Associates XD/RD-1000 Digital Transmission System*

The Math Associates, Inc., XD/RD-1000 is a complete, ready-to-operate optical-fiber transmission system designed to transmit a wide range of digital logic signals over distances of 1000 m or more.

The XD/RD-1000 system consists of the XD-1000 transmitter and RD-1000 receiver. Both units are fully compatible with most logic families, such as TTL, DTL, and CMOS, and all logic coding formats, including RZ, NRZ, and Manchester. The system is a pure, no-nonsense, dc-coupled, digital system and will handle any duty cycle or code format, limited only by rise and fall times. Speed of response is further limited to prevent extraneous pickup in electrically noisy areas. In addition, internal power supplies allow operation

_____
* This section is taken from Ref. 2.

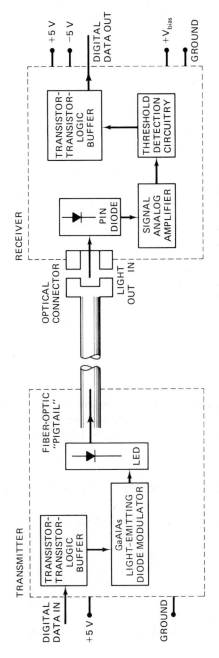

Figure 7.2 Block diagram of RCA fiber-optic link. (Courtesy of RCA.)

**TABLE 7.1   RCA Fiber-Optics Data Links
General Specifications**

	General
Output window	
Material (fiber optic cable)	
C86003E, C86004E, C86005E	DuPont PIFAX-120
	(200-μm core)
C86012E	Siecor 112
	(62.5-μm core, graded index)
Minimum length extending	
from package face	101 mm (4 in.)

Maximum Ratings, Absolute-Maximum Values		
Transmitter module		
power requirements		
Voltage (min.)	4.75	V
Voltage (max.)	5.25	V
Receiver module		
Power requirements		
Positive voltage	+5 ± 5%	V
Negative voltage	−5 ± 5%	V
Detector bias voltage	+45	V
Temperature (both modules)		
Storage	−35 to +75	°C
Operating	−35 to +50	°C

Source: Courtesy of RCA.

from batteries, unregulated dc sources or simple ac voltages, thereby providing unparalleled versatility in an optical-fiber transmission system.

Specifications for the system (shown in Fig. 7.3) are given in Table 7.3.

7.2 AN 8-KM FIBER-OPTIC CATV SUPERTRUNK SYSTEM*

The design of the first major fiber-optic digital cable TV (CATV) supertrunk in North America is described in the following paragraphs.

The supertrunk system transmits 12 NTSC color television channels and 12 FM stereo channels over an eight-fiber cable 7.8 km in length. The optical-fiber cable is lashed to messenger wires with optical repeaters at 2.8-km intervals. Audio, video, and FM signals are digitized and multiplexed into a single 322-Mb/s bit stream modulating an injection laser diode transmitter.

The system uses graded-index optical fibers having a bandwidth–distance product in excess of 600 MHz-km and attenuation less than 8 dB/km. Eight small optical fibers (0.005 in. in diameter) are formed into a

* This section is taken from Ref. 3.

TABLE 7.2 Typical Performance Characteristics for RCA Fiber-Optics Data Links

	C86003E	C86004E	C86005E	C86012E	
Transmitter module					
Power requirements					
Voltage	+5	+5	+5	+5	V
Current	250	250	250	250	mA
Input:					
digital signals—TTL-compatible					
NRZ coding	20	5	2	20	Mb/s
RZ coding	10	2.5	1	10	Mb/s
Output:					
min. peak optical power output	100	100	100	40	µW
Receiver module					
Power requirements					
Positive					
Voltage	+5	+5	+5	+5	V
Current	30	30	30	30	mA
Negative					
Voltage	−5	−5	−5	−5	V
Current	20	20	20	20	mA
Detector bias voltage	+5 to +45	+5 to +45	+5 to +45	+5 to +45	V
Input: optical sensitivity	0.50	0.25	0.10	0.50	µW
Output: digital signals—TTL-compatible					
NRZ coding	20	5	2	20	Mb/s
RZ coding	10	2.5	1	10	Mb/s
System data allowable loss	−23	−25	−27	−20	dB

Source: Courtesy of RCA.

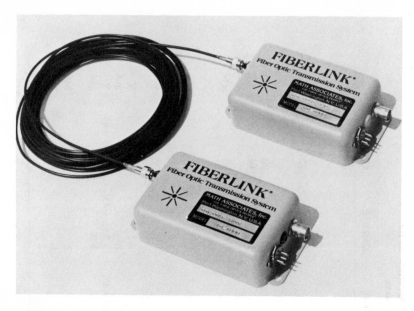

Figure 7.3 XD/RD-1000 digital transmission system. (Courtesy of Math Associates, Inc.)

**TABLE 7.3   XD/RD-1000 Digital Transmission System
Technical Specifications**

Complete system bandwidth	0–1 × 10⁶ pulses/s
System rise or fall time	0.2 μs or slower
Duty cycle	Limited only by RT and FT
Logical "1" input	2.0 V min., 5.5 V max.
Logical "0' input	0.8 V max., 0 V min.
Logical "1" output	2.4 V min., 3.3 V typ.
Logical "0" output	0.4 V max., 0.2 V typ.
Output current (logical "1")	30 mA max.
Allowable optical attenuation (10⁻⁸ B. E. R.)	− 30 dB typ.
Operating wavelength	890 nm (660 nm available)
Optical connectors	Amphenol 905 series
Power requirements	+ 7 to + 15 V dc at 200 mA
(transmitter and/or receiver)	or 6.3 to 14 V ac rms
	50/60 Hz at 200 mA unreg.
Physical size (per module)	6.4 × 12 × 3.2 cm

Source: Courtesy of Math Associates, Inc.

single multifiber cable providing transmission capacity for a full 12-TV-channel supertrunk in a cable less than ½ in. in diameter.

Of the eight fibers, only six are actively used, allowing the two spare fibers to be used for expansion to 18 channels. The six fibers are allocated as follows:

Fiber No.	Function
1,2	Each fiber carries three high-quality digital baseband TV channels, three digital FM stereo channels, plus parity and housekeeping data.
3,4,5	Each fiber carries two high-quality digital VSB TV channels, four digital FM stereo channels, plus housekeeping data.
6	Carries three channels of high-quality digital baseband in the opposite direction (i.e., from distribution hub to head end)

Thus, the system provides for full-duplex (simultaneous) two-way video communication.

The total system length selected for this system is 7.8 km, the distance between the head-end terminal location and the hub distribution location.

7.2.1 Baseband TV Processing

In order to transmit studio-quality signals, a baseband-encoded digital TV signal requires 8 bits sampled at or above the Nyquist rate. The sampling rate was chosen at 10.74 MHz, the third harmonic of the color subcarrier frequency. An additional ninth bit is required for multiplexing the audio portion of the signal and for carrying the frame synchronizing data. A tenth bit is used as a parity bit for error detection and concealment to reduce the effect of bit errors on decoded video quality. With this data format, each single baseband TV signal produces a 107.4-Mb/s digital bit stream.

Three such digital bit streams are then multiplexed into one serial bit stream at 322.2 Mb/s, which is in turn transmitted over the fiber-optic link. The FM stereo bits are interleaved into the bit stream into spare time-division-multiplexed time slots.

7.2.2 Vestigial Sideband Processing

The vestigial sideband (VSB) encoded TV signal is encoded in a similar manner to the baseband scheme previously described, but is sampled at a higher sampling rate of 16.1 MHz because of its higher analog bandwidth. The VSB digital data format produces a bit stream at 161.1 Mb/s. Two such bit streams are interleaved into one serial bit stream at 322.2 Mb/s for transmission over the link.

7.2.3 Fiber-Optic Link

The system uses injection laser diode (ILD) transmitters and avalanche photodiode (APD) receivers at both ends of the link as well as in repeaters.

A block diagram of the ILD transmitter is shown in Fig. 7.4. The transmitter outputs a 322-Mb/s NRZ digital code in optical form, giving over

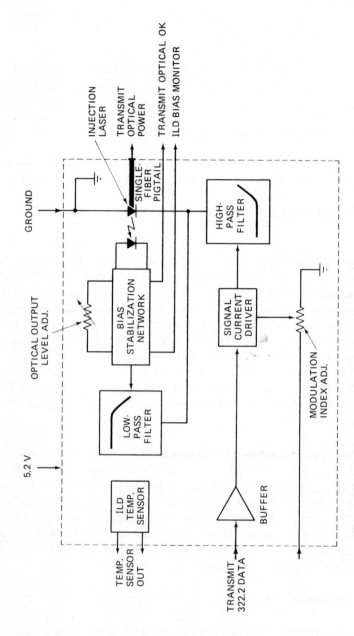

Figure 7.4 Injection laser transmitter, 322 Mb/s. (Courtesy of Harris Corp. and Cable-systems Engineering.)

1 mW of optical power into the fiber. An optical feedback circuit controls and stabilizes the laser threshold over temperature and time variations.

The APD receiver is shown in Fig. 7.5. The optical fiber stub terminates on the face of the photodiode, the output of which is amplified by a wideband, high-gain RF amplifier with AGC.

The fiber-optic link was designed to provide an overall end-to-end system BER of better than 1 error in 10^9 bits (BER $< 10^{-9}$) at a data rate of 322 Mb/s. Actual BER performance has been measured at better than 10^{-10}, which is the limit of the present test equipment.

Additional "safety margins" have been incorporated in the system design to ensure that the bit-error rate never limits system performance. For example, a parity detection circuit was designed into the system to search for bit errors in the four most significant bits of the digitized video sample. A logic processor stores the third previous video word to minimize changes in chrominance and luminance coherency, and the current word is checked for parity error. If an error is found, the previous correct sample is used, and the present erroneous sample is discarded. In effect, this technique of error detection and concealment allows errors to be located and removed prior to transmission to system subscribers. The result is a greater improvement in video service quality and reliability.

Another safety feature of the fiber-optic link is optical power margin, or link margin. This is the extra allowance made for equipment aging, variations in optical power, lossy fiber splices, and general time-variant system

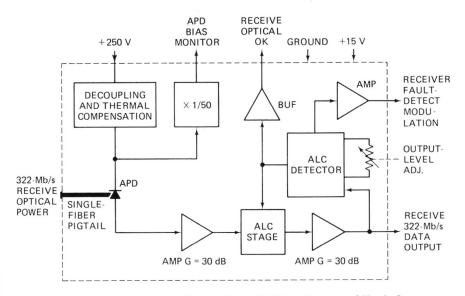

Figure 7.5 Avalanche photodiode receiver, 322 Mb/s. (Courtesy of Harris Corp. and Cablesystems Engineering.)

parameters. The link margin for this supertrunk system is 10 dB, which indicates that a significant optical link degradation can occur over time without the bit-error rate falling below the system BER specification of 10^{-9}.

REFERENCES

1. "Fiber Optics Data Links, Developmental Types C86003E, C86004E, C86005E, C86012E," RCA Electro Optics and Devices Publication.
2. "XD/RD-1000 Digital Transmission Set," Math Associates specification sheet.
3. Whitworth W. Cotten and C. Richard Patisaul, Harris Corp., and Donald G. Monteith, Cablesystems Engineering, "An 8 km Fiber-Optic CATV Supertrunk System," presented at the 27th Annual NCTA Convention, May 1, 1978.

8

MAINTENANCE

The installation and maintenance of fiber-optic systems obviously calls for different procedures than those used in most electronic systems. For instance, instead of concentrating on voltage and resistance measurements, the technician will be more concerned with measuring light intensity. Component standardization in fiber optics is a long way off, making interchangeability of parts difficult, if not impossible. Thus, when the technician finds a defective part, an exact replacement must be used.

In this chapter we discuss safety and fiber-optic measurements. Splicing and connectoring were discussed in Chapter 5.

8.1 SAFETY

Like any other new system, fiber-optic systems have some unexpected hazards which can cause personal injury if ignored. These dangers originate in the lasers and LEDs, the glass fibers and receiver power supplies, and in the materials and equipment used for installation and measurement. Specific precautions are given in the following paragraphs; general precautions may be found in the author's *Handbook of Electronic Safety Procedures* [1].

Chemical hazards. Freon and other solvents used in stripping and cleaning optical fibers may be hazardous. Avoid breathing vapors, especially if you are in a confined area such as a manhole [2]. When handling flame-retardant cables, avoid prolonged skin contact; wash hands before smoking or eating [3].

In some connectoring procedures, propanol is used as a lubricant. In such cases adequate ventilation is necessary because the fumes from the propanol are flammable [4]. Avoid contact with LOCTITE 495, an adhesive used in some cable preparation steps, as it cures instantly when in contact with human skin.

Electrical hazards.   In most circuits, only very low voltages (12 V or less) are in use. These voltages are harmless, of course, unless your feet are immersed in water at the bottom of a manhole. However, in some avalanche photodiode circuits in receivers, 300 V or more may be present. Such voltages should be treated with respect. Very high voltages are present in fusion splicers and in the oscilloscopes in optical time-domain reflectometers. Make sure that the frames of such equipment are well grounded.

Mechanical hazards.   Glass optical fibers can puncture the skin as well as the eyes. Use protective safety glasses when handling exposed fibers and during fiber cleaving, grinding, and polishing operations.

Radiation hazards.   It is very dangerous to stare at the end of an optical fiber that is illuminated by a laser or an LED [5]. These light sources can be hazardous whether they are part of the fiber-optic system or a part of the test equipment. Even though the radiation may be invisible in most cases, the retina of the eye can be damaged by looking at such radiation. Typical warning signs are shown in Fig. 8.1. Particular care should be taken to avoid viewing output flux under magnification.

8.2 MEASUREMENTS

Measurements are just as essential in the maintenance of fiber-optic systems as they are in strictly electronic systems. However, because of a lack of industry-wide standard measurement conditions, it is sometimes difficult to relate measurements by two different companies or organizations. Only when the test setups are identical can there be any assurance that test measurements are comparable.

Conventional measuring techniques can be followed in adjusting and troubleshooting the strictly electronic portions—receiver and transmitter—of a fiber-optic system. Therefore, we consider only *optical* measurements in this text.

Optical measurements require special test equipment, but simpler techniques can be used in an emergency. For example, if a fiber-optic cable has a metallic shield, shorts to ground can be checked with an ohmmeter. If there is

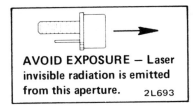

AVOID EXPOSURE — Laser
invisible radiation is emitted
from this aperture.   2L693

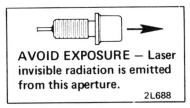

AVOID EXPOSURE — Laser
invisible radiation is emitted
from this aperture.
                    2L688

Figure 8.1   Typical laser warning labels.
(From Ref. 6; courtesy of RCA.)

a short to ground, this may indicate cable damage, which may imply fiber damage. Another simple test is to shine a flashlight into a fiber at one end and then look for light at the other end [7]. (*Careful:* Use of the eye as a sensor could lead to the bad habit of using the eye in all cases, even in hazardous situations.)

There are two basic types of measuring equipment for fiber optics: (1) optical power meters and (2) optical time-domain reflectometers. These instruments are discussed in the following paragraphs.

8.2.1 Optical Power Meters

Measurement principles.* The following measurements are necessary:

1. *Absolute radiant power output of the source,* which is as important to fiber optics as absolute current and voltage are to electronics. If an optical source delivers significantly less than its rated output, lowering the total dB loss of a system's passive components will not compensate for the weak input.

* This section is reprinted with permission from Ref. 8 (*Electronic Design,* Vol. 27, No. 21); copyright Hayden Publishing Co., Inc., 1979.

2. *Fiber power loss,* which depends on the fiber length and on the angle of launch; measurement of fiber loss can be difficult.

3. *Connector and splice losses,* which involve specifications of both the light coupled into a fiber's core and the light coupled into its cladding; fiber specs are usually for light coupled into the core only. The surest resolution to this inconsistency is to measure the optical power into and out of the connector or splice.

4. *Receiver sensitivity.* Whether or not the system photodetectors provide gain, they convert incident light into electrical current. Measuring the efficiency of the conversion—or responsivity—requires an optical-power meter and an ammeter.

It is most convenient to measure all these variables directly in decibels (dB)—the standard communications unit. When expressed in dB, system gains and losses can be easily perceived and quickly evaluated.

Absolute-power measurements in dBm and dBμ are important for evaluating active optical components, such as sources and receivers. Measurements of relative-dB loss are appropriate for passive optical components, such as fibers, connectors, splices, and tees. Some popular optical dB units are described in Table 8.1.

LEDs and injection laser diodes, the most popular fiber-optic emitters, have small active areas; their total light-output power, therefore, is readily measurable by a large-area photodiode detector placed nearby, as the setup in Fig. 8.2 shows.

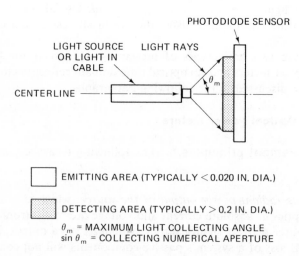

Figure 8.2   Active areas of fiber-optics emitters. Fiber-optic emitters have small active areas, so their total light-output power is readily measurable with a large-area detector placed near the source. [Reprinted with permission from Ref. 8 (*Electronic Design,* Vol. 27, No. 21); copyright Hayden Publishing Co., Inc., 1979.]

TABLE 8.1   Enlightment on Optical dB Units

Measurements of optical radiation power are expressed in watts. Decibel (dB) power is

$$dB = 10 \log \left(\frac{P_{sig}}{P_{ref}} \right)$$

where P_{sig} is the power to be measured and P_{ref} is the reference power.
For 1-mW reference power:

$$dB_m = 10 \log \left(\frac{P_{sig}}{1 \text{ mW}} \right)$$

For 1-μW reference power:

$$dB\mu = 10 \log \left(\frac{P_{sig}}{1 \text{ } \mu\text{W}} \right)$$

With both P_{sig} and P_{ref} variable, the dB-power formula expresses the log ratio of the two unknowns in dB.
Light-power loss of an optical data-link element is

$$L \text{ (dB)} = 10 \log \left(\frac{P_o}{P_{in}} \right)$$

The input power (P_{in}) and output power (P_o) of a component can be measured in dBm units, dBμ units, or dB units without a known reference. The loss, L, expressed in decibels, is the same:

$$L \text{ (dB)} = dBm \text{ } (P_o) - dBm \text{ } (P_{in})$$
$$= dB\mu \text{ } (P_o) - dB\mu \text{ } (P_{in})$$

Source: Reprinted with permission from Ref. 8 (*Electronic Design,* Vol. 27, No. 21); copyright Hayden Publishing Co., Inc., 1979.

Photodiode sensors with areas of only 1 cm^2 can handle half-cone acceptance angles (θ_m) greater than 45 degrees. This acceptance angle corresponds to a numerical aperture for light collection (sin θ_m) of more than 0.7, which, of course, is quite high.

The numerical aperture is also a measure of the angular distribution of power emitted from the source. This distribution can be tested by scanning a pinhole mask across the detector's field of view of the source.

Sources for fiber-optic communications systems usually emit in the spectral range 700 to 1400 nm. The spectral response of silicon photodiodes matches the range from 700 to 1100 nm; germanium devices can measure sources from 1000 up to 1400 nm.

Silicon detectors are more sensitive than germanium detectors and can measure lower power. A 1-cm^2 silicon photodiode can measure absolute power down to 1 pW. Higher leakage currents and lower internal impedances limit germanium diodes to 10-nW measurements. However, germanium diodes, unlike silicon photodiodes, can measure wavelengths longer than 1100 nm.

The power output of the source is one of the factors involved in the calculation of fiber loss, which is most conveniently represented by the cable-loss factor (CLF) in dB/km:

$$CLF = \frac{P_i - P_o}{L}$$

where

P_i = input power, dB

P_o = output power, dB

L = cable length, km

The power output of the source is easier to measure than the effective input power to the fiber. Loss mechanisms extract a heavy toll on light sent to the fiber's core; for example, not all of the source light falls on the core and the launch angle may direct some of the light into the fiber cladding.

The difference between the optical power input and the optical power output of a connector or splice equals either the connector loss or the splice loss. Making measurements directly in dB simplifies the required calculations. If the connector in question is attached to a long cable, the CLF must be subtracted from the total loss measured for the connector.

The overall measure of power losses from a system may not correlate to the manufacturer's specifications for losses from individual components. Some makers of connectors specify the loss both for light coupled into the fiber core and for light coupled into the cladding. However, cable specifications often include only the light coupled into the core. In addition to the losses specified for connectors and cable, an accurate appraisal of total-link loss requires measurement of the loss into fiber cladding at a connector interface.

The quality of the conversion of input optical power into output electrical current is called responsivity (measured in A/W). To make this conversion, photoreceivers for fiber optics most frequently employ photodiodes (composed of Si, GaAlAs, or Ge) as the basic photosensor. Photodiode signals are amplified and processed according to the requirements of the system. The photodiodes can be PIN types, without internal gain, or avalanche types, with internal gain. The responsivity of the photodiodes is usually linear but varies with wavelength, which must therefore be measured at the source.

Both an optical power meter and a current meter are needed to measure photodiode responsivity. First, the optical power emitted from a stable source at the desired system wavelength is measured with the power meter. The calibrated source excites the current from the photodiode. This current is then measured by the current meter. The ratio of optical power to excited current is the photodiode responsivity.

The photodiode responsivity (A/W) and the conversion factor (V/A) for the amplifier following the photodiode equal the overall receiver transfer function (in volts out per watt of optical power in). The receiver transfer function, together with the output noise voltage of the photoreceiver (measured with the optical power off), determines the system's minimum ability to detect light.

Typical test equipment.   The following paragraphs described some of the test equipment available to make the measurements just discussed.

Photodyne's Model 22XL optical multimeter (Fig. 8.3) measures light sources, photoreceivers, fiber-cable transmission, and connector and splice loss. It gives a direct power readout in dBμ or dBm over a power range of -90 to $+30$ dBm for a spectral range of 220 to 2000 nm. Interchangeable sensor heads extend the capabilities of this unit. Adapters are available for most common fiber-optic connectors. As a lightweight, hand-held meter, it is convenient for field use.

The Fiberguide OP-3 fiber-optics meter (Fig. 8.4) measures LED and laser optical power and emittance angles, optical-fiber and connector loss, and optical-fiber numerical aperture. It has a spectral range of 600 to 1050 nm and an optical power range between 2000 μW and 20 nW($+3$ to -47 dBm).

The Fiberguide OM1 optical ohmmeter (Fig. 8.5) measures fiber loss and provides a digital readout in decibels. In operation, a standard reference cable is used to calibrate the instrument. Then the unknown cable is attached to the optical connectors. It operates at a wavelength of 820 nm with a

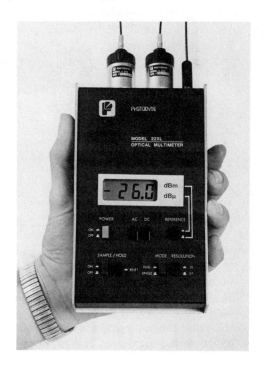

Figure 8.3 Photodyne Model 22XL optical multimeter. (Courtesy of Photodyne.)

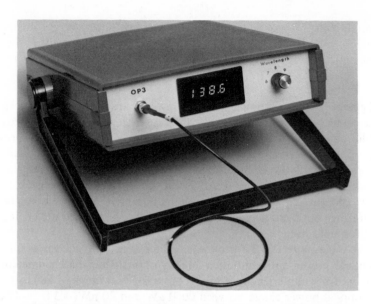

Figure 8.4 Fiberguide OP-3 power meter. (Courtesy of Fiberguide.)

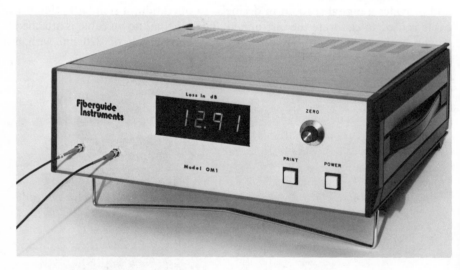

Figure 8.5 Fiberguide OM1 optical ohmmeter. (Courtesy of Fiberguide.)

measurement range of $+5$ to -50 dB. The measurement resolution is 0.01 dB on all ranges.

Fiberguide's OM-2 fiber attenuation testing system (Fig. 8.6) is designed for high-throughput production applications. Plug-in modules enable the user to change the numerical aperture, spot size, and wavelength. It measures up to 55 dB with a maximum resolution of 0.01 dB.

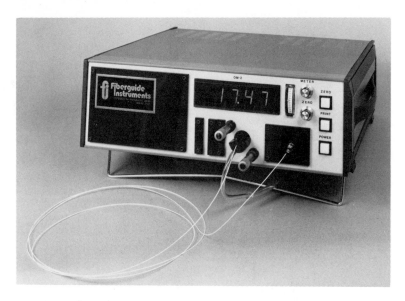

Figure 8.6 Fiberguide OM–2 fiber attenuation testing system. (Courtesy of Fiberguide.)

Times Fiber's Model 125 fiber attenuation measurement system measures fibers with losses up to approximately 40 dB and numerical apertures up to 0.5. Wavelength is continuously variable between 700 and 1600 nm. Readout is with a dc picoammeter.

AEG-Telefunken's Model SP 773 attenuation test set measures attenuation of fibers with core diameters greater than 42 μm. It has a range up to 60 dB at a wavelength of 820 nm. A block diagram of the set is shown in Fig. 8.7.

The Hamamatsu C-1308 Picosecond light pulser (Fig. 8.8) is useful in testing fast photodetectors and in impulse response measurements of fiber cables. It uses a laser diode to generate a light pulse centered around 800 nm with a spectral bandwidth of approximately 3 nm and a pulse width duration of 100 ps FWHM (typical).

The Hewlett-Packard 84801A thermistor sensing device (Fig. 8.9) is an optical power sensor dedicated to fiber optics. Used in conjunction with any of the HP 432 series power meters, it measures absolute optical power from − 30 dBm to + 10 dBm (1 μW to 10 mW) over the spectral range 600 to 1200 nm.

In use, light from the fiber under test or another optical power source is coupled into the input fiber, and the sensor's output is connected to the power meter. The 84801A/432 combination measures directly the incident optical power coupled into the fiber at the input, and displays in watts or dBm. The combination measures the output power from optical sources such as light-emitting diodes or lasers which have a pigtail fiber attached to them. It is also

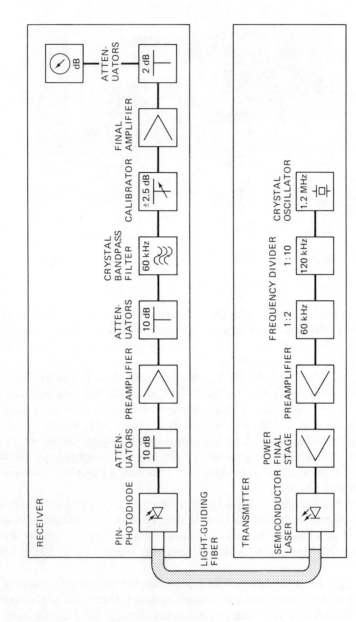

Figure 8.7 Block diagram of AEG-Telefunken attenuation test set. (Courtesy of AEG-Telefunken.)

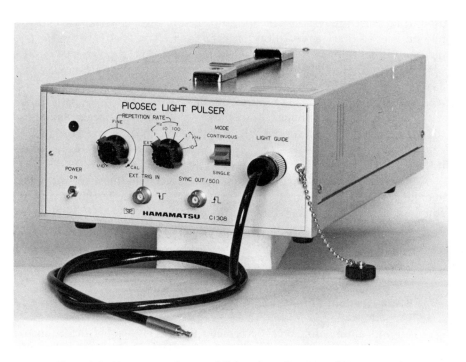

Figure 8.8 Hamamatsu picosecond light pulser. (Courtesy of Hamamatsu.)

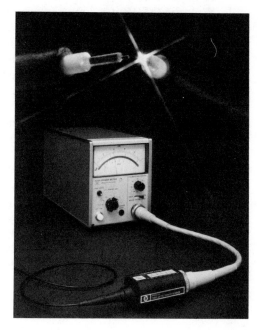

Figure 8.9 HP 84801A fiber-optic power sensor with HP 432A power meter. (Courtesy of Hewlett-Packard.)

used to measure the response of optical detectors such as avalanche and PIN photodiodes to find the deviation from the square law, that is, when the output current is no longer proportional to the input optical power.

Because of the lack of optical connector standardization, input to the 84801A is made through a 1-m length of 200-μm-core-diameter, 0.4-NA plastic-clad silica fiber. Alignment of this fiber to the test fiber can be made with user-supplied positioners or connectors. Several connector manufacturers, including AMP, Amphenol, and Hughes, make connectors directly compatible with the 84801A fiber; other connectors can be modified to accept it.

The fiber of the system under test can be precisely matched, with index matching fluid, to the input fiber of the sensor, thereby improving the coupling efficiency and thus the measurement accuracy. The fluid also protects the fiber edges from scratches.

Relative power measurements, within the dynamic range of the system, can be made to accuracies better than $\pm 2\%$.

8.2.2 Optical Time-Domain Reflectometers*

An optical time-domain reflectometer (OTDR) is a sophisticated instrument designed to (1) determine the length of optical fibers without physically measuring them; (2) locate breaks or fractures in a fiber; (3) measure the attenuation of fibers, splices, and connectors; and (4) determine the distance to discontinuities such as splices and connectors.

With the OTDR it is not necessary to cut the fiber or to have access to both ends of the fiber in order to make measurements. Because of these features, the OTDR is ideal for testing and troubleshooting fiber-optic systems, both in the laboratory and in the field.

As shown in Fig. 8.10, the OTDR consists of a laser pulser, a coupler, a detector, a processor, and an oscilloscope. The oscilloscope is essential for the OTDR but, depending on the manufacturer, it may or may not be an integral part of the OTDR. Figures 8.11, 8.12, and 8.13 show typical OTDRs.

The OTDR uses the same principle of measuring as the conventional electronic time-domain reflectometer. In a typical OTDR, the laser emits a series of short (10 to 100 ns), intense (0.5 W) pulses at 820 nm. The pulses pass through a collimating lens, a 3-dB beam splitter, and a second lens into the end of the fiber to be checked.

When a pulse hits any discontinuity—break, splice, connector, and so on—a small quantity of it is reflected back toward the source. The beam splitter couples some of this reflection to an avalanche photodiode (APD). The signal detected by the APD is processed and sent to an oscilloscope for

---

* This section is taken from Refs. 9 and 10.

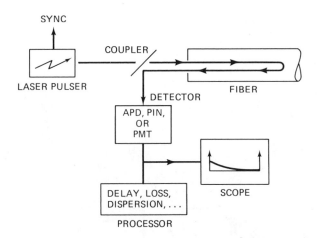

Figure 8.10 Optical time-domain reflectometer block diagram. (From Ref. 11; copyright 1978 IEEE.)

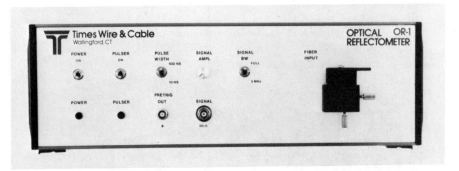

Figure 8.11   Times Wire & Cable OTDR. (Courtesy of Time Wire & Cable.)

display. Interpretation of the oscilloscope traces allows optical parameters to be determined.

Attenuation can be determined from the slope of the backscattering curve. Splice and connector losses can be found by evaluating the offsets in this curve. Fiber length can be calculated from the time difference between the reflected pulses from the front and far ends of the fiber.

It is much easier to determine the length of a cable on a large spool with an OTDR than it is to unroll the spool and physically measure the cable.

The OTDR provides an excellent method of pinpointing faults or breaks in a fiber. Maximum length that can be tested varies from 5 to 10 km, depending on the manufacturer and the cable loss. Accuracy of fault location varies from ± 1 to ± 2 m with these models.

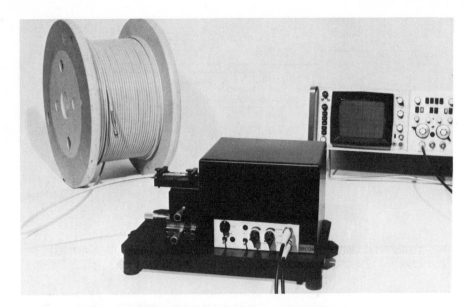

Figure 8.12   Siecor OTDR (Courtesy of Siecor.)

Figure 8.13   ITT OTDR (Courtesy of ITT.)

The output of the OTDR can generally be connected to a chart recorder in order to obtain a permanent copy. At least one manufacturer offers an OTDR with a digital read-out in addition to the oscilloscope. Because of the narrow pulses, the oscilloscope must be a wideband instrument, generally 50 to 80 MHz.

During a measurement it is recommended that the fiber's end be cleaved flat. A drop of coupling fluid should then be placed against the OTDR optical port before positioning the fiber.

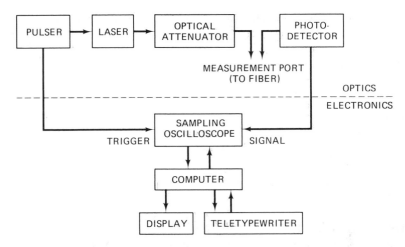

Figure 8.14   Delay-distortion measurement test set (after Dannwolf). (From Ref. 11; copyright 1978 IEEE.)

Table 8.2 gives specifications for a typical OTDR.

Delay distortion can be measured by impulse-response measurements through the test setup shown in Fig. 8.14. The pulse spreading is measured directly in the time domain and the results are converted to the equivalent fre-

TABLE 8.2   **Times Wire & Cable OTDR OR-1 Specifications**

Maximum length	40 dB of optical loss
Input pulse characteristics	
Wavelength	800 nm
Pulse width	10 and 100 ns
Repetition rate (pulses/s)	100 to 700 for 80 ns
	100 to 10,000 for 10 ns
Peak power	200 mW
Output signal characteristics	
Amplitude	200 mV
Pulse width	15 ns/85 ns (50%)
Background noise	200 μV
Impedance	50 Ω
Acceptable fiber diameters	
Standard fixture	150 μm, max.
Larger diameter fixtures	Optional
Bench mounted	
Size ($W \times H \times D$)	17 in. (43 cm) $\times$ 5 in. (12.5 cm) $\times$ 13 in. (33 cm)
Weight	15 lb (6.8 kg)
Power requirements	
supply	120 V ac, 60 Hz, 2 A

Source: Courtesy of Times Wire & Cable.

quency response. Another technique is to measure the response directly in the frequency domain [12].

REFERENCES

1. E. Lacy, *Handbook of Electronic Safety Procedures* (Englewood Cliffs, N. J.: Prentice-Hall, Inc., 1977).
2. Amphenol RF Division Specification 349-502-3, Nov. 5, 1976.
3. Du Pont Safety Instruction, n.d.
4. "HFBR-0100, Fiber Optic Connector Assembly Tooling Kit," Hewlett-Packard User's Manual, Nov. 1980.
5. C. C. Timmermann, "Handling Optical Cables: Safety Aspects," *Applied Optics,* Vol. 16, No. 9, Sept. 1977, pp. 2380-2382.
6. "Optical Communications Products," RCA Publication OPT-115, June 1979.
7. J. Adams, C. Swinn, and G. Kavanagh, "Field Testing Fiber Optics," *Telephony,* May 14, 1979, pp. 29-31.
8. Paul Wendland, "Lighten the Burden of Fiber-Optic Measurements with New Instruments, Standards," *Electronic Design,* Vol. 27, No. 21, Oct. 11, 1979, pp. 126-130.
9. "OTDR Model OR-1 Preliminary Operating Instructions," Times Fiber Communications, n.d.
10. "Optical Time Domain Reflectometer," Siecor, n.d.
11. Michael K. Barnoski and S. D. Personick, "Measurements in Fiber Optics," *Proceedings of the IEEE,* Vol. 66, No. 4, Apr. 1978, pp. 429-440.
12. David Charlton and Paul R. Reitz, "Making Fiber Measurements," *Laser Focus,* Sept. 1979, pp. 52-64.

9

INTEGRATED OPTOELECTRONICS

Integrated optoelectronics is the technology of placing numerous microscopic optical and electronic devices on a single plane on a common chip. These devices typically are lasers, light waveguides, modulators, couplers, photodetectors, and transistors. Basic to all of them is a thin transparent-film waveguide for transmitting light energy.

Although practical integrated optoelectronic circuits may not be available commercially for a few more years, it is important that you understand the basics of these exciting new circuits.

Just as integrated circuits have rebuilt and shaped the future of electronics, integrated optoelectronics is expected to have an equally major impact on fiber optics. The expected advantages are considerable:

1. Lower drive-power requirements
2. Compact (microminiaturization)
3. Inexpensive
4. Faster
5. Very large bandwidth (high data-handling capacity)
6. Immunity to electromagnetic interference
7. Rugged
8. Stable
9. Efficient

With these advantages there can be numerous applications of these circuits, such as in data and signal processing. But the biggest application is expected to be in fiber-optic communications.

Construction of integrated optoelectronic circuits will make use of many of the techniques used in the manufacture of integrated electronic circuits. The big difference is that the optical circuits are much smaller, requiring manufacturing tolerances of less than a thousandth of a millimeter.

Fabrication of miniature thin-film optical components such as couplers and lasers has been easier than the manufacture of a *complete* integrated optoelectronic circuit on a single substrate. *Hybrid* circuits have been built with a combination of discrete components and integrated circuits. The goal, however, is to build a complete monolithic integrated optoelectronic circuit on one chip. This circuit would contain several devices which would generate, couple, modulate, and detect light.

Figure 9.1 shows some typical optical building blocks, most of which are described in the following paragraphs. The thin-film laser shown has an unacceptably short life for the present; therefore, a discrete laser is used in its place.

9.1 BEAM-TO-FILM COUPLERS

In theory, a laser beam can be coupled into a thin-film waveguide simply by focusing the beam on the end face of the waveguide. Unfortunately, in this case its efficiency is unreasonably low.

Laser beams can, however, be fed into a thin-film waveguide by either a thin-film prism coupler, a grating coupler, or a tapered film (tapered waveguide) coupler, shown in Figs. 9.2 and 9.3. Note that these couplers can be rearranged and used as ouput devices as well as for input. The prism coupler was the first to be developed and is therefore more widely used. It is a versatile unit but not as useful for integrated optoelectronics as are the grating coupler and the tapered film coupler. All three types have a coupling efficiency of approximately 80%.

The thin-film prism coupler is basically a thick section of the waveguide; it is about 1 mm tall and several micrometers thick.

9.2 WAVEGUIDES*

A thin-film of zinc sulfide 1 μm thick can serve as a light waveguide if it has a higher refractive index than those of surrounding materials. Within this waveguide, light rays are reflected through total internal reflection and follow a zigzag path similar to that in step-index optical fibers.

---

* This section is taken from Refs. 1, 3, and 4.

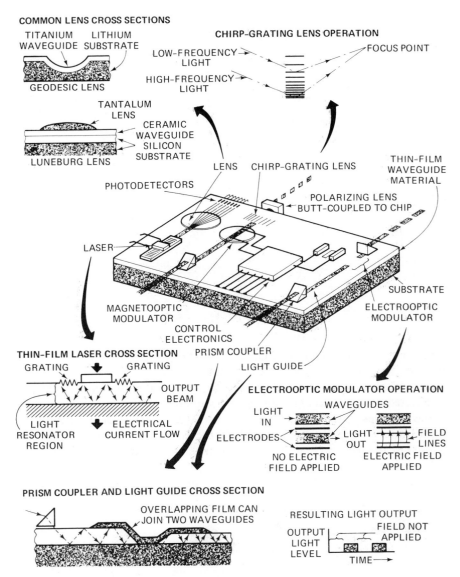

COMMON LENS CROSS SECTIONS

TITANIUM   LITHIUM
WAVEGUIDE SUBSTRATE

GEODESIC LENS

TANTALUM
LENS
CERAMIC
WAVEGUIDE
SILICON
SUBSTRATE
LUNEBURG LENS

CHIRP-GRATING LENS OPERATION

LOW-FREQUENCY
LIGHT

HIGH-FREQUENCY
LIGHT

FOCUS POINT

PHOTODETECTORS

LENS   CHIRP-GRATING LENS

THIN-FILM
WAVEGUIDE
MATERIAL

POLARIZING LENS
BUTT-COUPLED TO CHIP

LASER

SUBSTRATE

MAGNETOOPTIC
MODULATOR

CONTROL
ELECTRONICS

ELECTROOPTIC
MODULATOR

THIN-FILM LASER CROSS SECTION   PRISM COUPLER

GRATING   GRATING   LIGHT GUIDE

OUTPUT
BEAM

LIGHT
RESONATOR
REGION

ELECTRICAL
CURRENT FLOW

ELECTROOPTIC MODULATOR OPERATION

WAVEGUIDES

LIGHT
IN

ELECTRODES

LIGHT
OUT

FIELD
LINES

NO ELECTRIC
FIELD APPLIED

ELECTRIC FIELD
APPLIED

PRISM COUPLER AND LIGHT GUIDE CROSS SECTION

OVERLAPPING FILM CAN
JOIN TWO WAVEGUIDES

RESULTING LIGHT OUTPUT

OUTPUT
LIGHT
LEVEL

FIELD NOT
APPLIED

TIME →

Figure 9.1   Typical optical building blocks. (From Ref. 1; copyright 1979 Machine Design.)

If the light encounters rough spots on the inner surfaces of the film it will scatter. But with a very smooth film these scattering losses can be made tolerable.

Many modes can be transmitted without interference to each other.

The forward velocity of light in the film depends on the thickness of the film. A thicker film will slow the light down. Thin and thick films may be

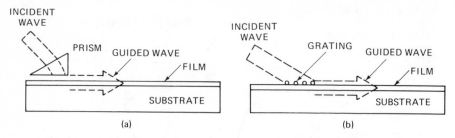

Figure 9.2 Prism and grating couplers: (a) prism coupler; (b) grating coupler. (From Ref. 2; copyright 1972 IEEE.)

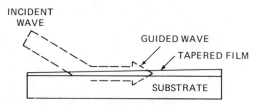

Figure 9.3 Tapered film coupler.

joined, but the junction must be tapered to prevent substantial losses. Typical thin-film junctions are shown in Fig. 9.4.

By placing one or more corrugated sections in the waveguide, light can be reflected or refracted. The corrugated section consists of precisely spaced parallel grooves. The distance between the ridges is called the period of the corrugation. If this distance is one-half the wavelength (or some multiple thereof) of the light wave to be reflected, the corrugated section will act as a mirror for that particular wavelength.

This phenomenon can be put to use in a light band-rejection filter.

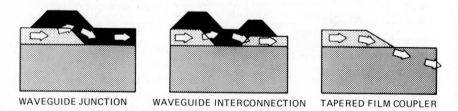

WAVEGUIDE JUNCTION   WAVEGUIDE INTERCONNECTION   TAPERED FILM COUPLER

Figure 9.4 Thin-film junctions. A beam of light (arrow) can be made to pass from one thin-film device to another simply by overlapping the tapered ends of both films (left). Alternatively (center), the tapered devices can be positioned end to end, with both ends and the intervening space coated with a material whose refractive index is greater than that of the substrate. Or the light wave can be made to pass into the lower-refractive-index substrate (right). (From Ref. 5; copyright 1980 Bell Laboratories RECORD.)

Another application allows separation of different wavelengths traveling in the same waveguide. This is a useful feature for wavelength multiplexing, discussed in Chapter 3.

9.3 LASERS

A semiconductor diode laser for integrated optoelectronics can be made of a layer of gallium arsenide inserted between layers of gallium–aluminum arsenide. Instead of cleaved end-face mirrors used in the injection lasers we considered in Chapter 3, corrugated sections are used to reflect the light [3]. The laser output beam is emitted when the light gets up enough energy to pass through one of these sections.

Other types of lasers are being developed, as we will see in Section 9.6.

9.4 MODULATORS

In previous chapters we have pointed out that discrete injection lasers and LEDs are generally modulated by varying the diode current. However, in the early development of optoelectronic circuits, this modulation technique could not be used because of the tendency of the lasers to oscillate. Instead, external modulators had to be used.

Such modulators are made of a material whose refractive index can be changed by an applied electric field (as in electro-optic modulators) or magnetic field (as in magneto-optic modulators). Electro-optic modulators produce phase modulation of the light; magneto-optic modulators give amplitude modulation.

One common type of electro-optic modulator is the optical directional coupler. The coupler consists of two transparent thin-film strips designated waveguide A and waveguide B. The strips are placed very close together. Electrodes are placed on two sides of waveguide B, the output device.

If there is no power applied to these electrodes, light from waveguide A will leak into waveguide B. All the power in waveguide A is coupled into waveguide B. However, if power is applied to the electrodes, an electric field will be placed across waveguide B. This field stops the interchange of power and in effect makes the directional coupler a light switch. Thus, depending on the applied voltage, light emerges either from waveguide A or waveguide B.

Although this switching characteristic is useful in optical distribution networks, it is also useful for modulating an optical beam. If a stream of binary pulses is applied to the electrodes while a continuous laser beam is applied to the input of waveguide A, the power out of waveguide A will be modulated.

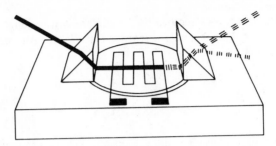

Figure 9.5   Magneto-optic switch. Incoming light from the left is polarized by a magnetic field around the serpentine electric circuit. A prism (right) can separate the light into one of two paths according to its direction of polarization. (From Ref. 6; copyright 1980 Bell Laboratories RECORD.)

A magneto-optic switch which is used as a modulator is shown in Fig. 9.5. A polarizing filter is placed at the output to allow only one beam to proceed.

With recent improvements in integrated optoelectronic circuits, external modulators may be rejected in favor of current modulators.

9.5 PHOTODETECTORS AND OTHER DEVICES

Figure 9.6 shows an early model of an integrated optical photodetector.

High-frequency operation of a laser integrated with an FET has been accomplished with the circuit shown in Fig. 9.7.

IBM researchers have developed an integrated fiber-optic transmitter (Fig. 9.8) which contains a semiconductor laser array, a cylindrical lens, and an array of optical fiber light guides. The components are mounted on a silicon wafer which also contains thin-film drive electrodes for the lasers.

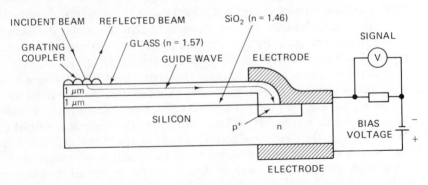

Figure 9.6   Integrated optical photodetector. (From D. B. Ostrowsky, "Optical Waveguide Components," in *Optical Fibre Communications,* ed. M. J. Howes and D. V. Morgan; copyright 1980 John Wiley & Sons Ltd.; reprinted by permission of John Wiley & Sons Ltd.)

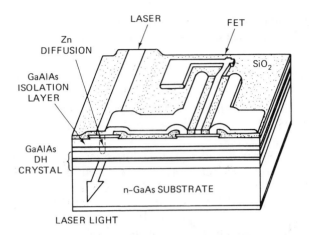

Figure 9.7 Integrated laser and FET. Hitachi's device in which a laser and FET are integrated on a single GaAs substrate. (From Ref. 7; copyright 1980 Advanced Technology Publications.)

The laser arrays contain up to 13 lasers fabricated in a single bar of gallium–arsenide–gallium aluminum arsenide double-heterojunction material. Each laser has an output power of up to 50 mW in continuous operation. Up to 70% of this power is coupled into the optical fibers.

The cylindrical lens used to focus the output light from the lasers into

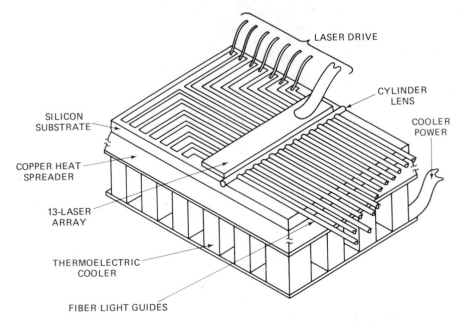

Figure 9.8 IBM's integrated fiber-optic transmitter. (Courtesy of IBM.)

the fiber light guides is a short piece of glass fiber with a diameter of 70 μm. The lightguides are made of low-loss graded-index fiber.

9.6 REPEATER

An experimental repeater has been developed which has a complete optoelectronic circuit [8]. The structure and equivalent schematic diagram of the circuit are shown in Fig. 9.9.

The laser shown is called a crowding-effect laser. Unlike earlier integrated optoelectronic circuits, this laser is current modulated and does not use external modulation. Transistor Q2 is the photodetector; its active load is transistor Q1. Transistor Q3 modulates the current that passes through the laser diode. The optical power gain of the repeater is 10 dB. Unfortunately, this is not a practical circuit. Because of the limitations of the crowding-effect laser, the circuit cannot operate continuously. By replacing this laser with a beryllium implanted laser, it may be possible to have a useful circuit.

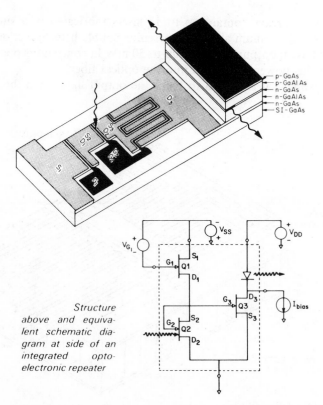

Structure above and equivalent schematic diagram at side of an integrated optoelectronic repeater

Figure 9.9 Integrated optoelectronic repeater. (From Ref. 8; copyright 1980 Advanced Technology Publications.)

REFERENCES

1. Leland Teschler, "The Promise of Integrated Optics," *Machine Design,* Dec. 6, 1979.
2. Stewart E. Miller, "A Survey of Integrated Optics," *IEEE Journal of Quantum Electronics,* Vol. QE-8, Part 2, Feb. 1972, pp. 199-206.
3. Amnon Yariv, "Guided-Wave Optics," *Scientific American,* Vol. 240, No. 1, Jan. 1979, pp. 64-72.
4. P. K. Tien, "Integrated Optics," *Scientific American,* Apr. 1974.
5. Ping-King Tien and Joseph A. Giordmaine, "Integrated Optics; Wave of the Future," *Bell Laboratories RECORD,* Dec. 1980, pp. 371-378.
6. Ping-King Tien and Joseph A. Giordmaine, "Integrated Optics: the Components," *Bell Laboratories RECORD,* Jan. 1981, pp. 8-13.
7. "3. Caltech Effort Begins to Integrate Optical and Electronic Components," *Laser Focus,* May 1980, p. 44.
8. Schlomo Margalit and others, "Integrated Optoelectronics," *Laser Focus,* Sept. 1980, pp. 76-80.

GLOSSARY*

Acceptance Angle: The angle measured from the longitudinal centerline up to the maximum acceptance angle of an incident ray that will be accepted for transmission along a fiber. The maximum acceptance angle is dependent on the indices of refraction of the two media that determine the critical angle. For a cladded glass fiber in air, the sine of the maximum acceptance angle is given by the square root of the difference of the squares of the indices of refraction of the fiber core glass and the cladding. *See* Maximum acceptance angle.

Acceptance Cone: A cone whose included apex angle is equal to twice the acceptance angle.

Analog-Intensity Modulation: In an optical modulator, the variation of the intensity (i.e., instantaneous output power level) of a light source in accordance with an intelligence-bearing signal or continuous wave, the resulting envelope normally being detectable at the other end of a light-wave transmission system.

Aperture: In an optical system, an opening or hole, through which light or matter may pass that is equal to the diameter of the largest entering beam of

* Adapted from "Vocabulary for Fiber Optics and Lightwave Communications," National Communication System (NCS), Technical Information Bulletin 79-1, Feb. 1979, prepared by the NCS Office of Technology and Standards. Deletions are indicated by ellipses. Changes and additions are indicated in brackets. References to "frequencies" have been changed to "wavelengths," and references to "microns" have been changed to "nanometers" and the units changed accordingly.

light that can travel completely through the system and that may or may not be equal to the aperture of the objective. *See* Numerical aperture.

Avalanche Photodiode (APD): A photo-detecting diode that is sensitive to incident photo energy by increasing its conductivity by exponentially increasing the number of electrons in its conduction-band energy levels through the absorption of the photons of energy, electron interaction, and an applied bias voltage. The photodiode is designed to take advantage of avalanche multiplication of photocurrent. As the reverse-bias voltage approaches the breakdown voltage, hole–electron pairs created by absorbed photons acquire sufficient energy to create additional hole–electron pairs when they collide with substrate atoms. Thus, a multiplication effect is achieved.

Blackbody: An ideal body that would absorb all radiation incident on it. When heated by external means, the spectral energy distribution of radiated energy would follow curves shown on optical spectrum charts. The ideal blackbody is a perfectly absorbing body. It reflects none of the energy that may be incident upon it. It radiates (perfectly) at a rate expressed by the Stefan–Boltzmann law and the spectral distribution of radiation is expressed by Planck's radiation formula. When in thermal equilibrium, an ideal blackbody absorbs perfectly and radiates perfectly at the same rate. The radiation will be just equal to absorption if thermal equilibrium is to be maintained. *Synonym:* ideal blackbody.

Brightness: An attribute or visual perception in accordance with which a source appears to emit more or less light. Since the eye is not equally sensitive to all colors, brightness cannot be a quantitative term. It is used in nonquantitative statements with reference to sensations and perceptions of light. . . .

Bundle: A group of . . . optical fibers . . . associated together and usually in a single sheath. Official fiber bundles are usually considered to be in a random arrangement and are used or considered as a single transmission medium. *See* . . . Coherent bundle; Optical fiber bundle. . . .

Cable: A jacketed bundle or jacketed fiber, in a form that can be terminated. *See* . . . Fiber-optic cable. . . .

Cable Assembly: A cable terminated and ready for installation. . . .

Cable Core: The portion of a cable inside a common covering.

Cable Jacket: The outer protective covering applied over the internal cable elements.

Candela: The luminous intensity of 1/600,000 of a square meter of a blackbody radiator at the temperature of solidification of platinum, 2045 Kelvin. [A] 1-candela [point source] emits 4π lumens of light flux.

Candlepower: A unit of measure of the illuminating power of any light source, equal to the number of [candelas] of the source of light. A flux density of 1 lumen of luminous flux per steradian of solid angle measured from the source is produced by a point source of 1 candela emitting equally in all directions.

Chromatic Dispersion: Dispersion or distortion of a pulse in an optical waveguide due to differences in wave velocity caused by variations in the indices of refraction for different portions of the guide. . . .

Cladding: An optical conductive material with a lower refractive index placed over or outside the core material of . . . an optical fiber, or a thin film on a substrate, that serves to reflect or refract light waves so as to confine them to the core, and serves to protect the core. . . .

Coherent Bundle: A bundle of optical fibers in which the spatial coordinates of each fiber are the same or bear the same spatial relationship to each other at the two ends of the bundle. *Synonym:* aligned bundle.

Coherent Light: Light that has the property that at any point in time or space, particularly over an area in a plane perpendicular to the direction of propagation or over time at a particular point in space, all the parameters of the wave are predictable and are correlated. . . .

Converging Lens: A lens that adds convergence to an incident bundle of light rays. One surface of a converging lens may be [convex] and the other plane planoconvex. Both may be convex (double-convex, biconvex) or one surface may be convex and the other concave (converging meniscus). *Synonyms:* convergent lens; convex lens. . . .

Core: The central primary light-conducting region of . . . an optical fiber, the refractive index of which must be higher than that of the cladding in order for the light waves to be internally reflected or refracted. Most of the optical power is in the core. *See also* Cladding; Cable core; Fiber core.

Coupler: In optical transmission systems, a component used to interconnect three or more optical conductors. *See* . . . Data bus coupler; Non-reflective . . . coupler; Reflective star coupler; Tee coupler. . . .

Critical Angle: In terms of indices of refraction, the critical angle is the angle of incidence from a denser medium at an interface between the denser and less dense medium, at which all of the light is refracted along the interface (i.e., the angle of refraction is 90°). When the critical angle is exceeded, the light is totally reflected back into the denser medium. The critical angle varies with the indices of refraction of the two media with the relationship $\sin A\ (C) = n(2)/n(1)$, where $n(2)$ is the index of refraction of the less dense medium, $n(1)$ is the index of refraction of the denser medium, and $A(C)$ is the critical angle, as above. In terms of total internal reflection in an optical fiber, the critical angle is the smallest angle made by a meridional ray in an optical fiber that can be totally reflected from the innermost interface and thus determines the maximum acceptance angle at which a meridional ray can be accepted for transmission along a fiber. *See also* Total internal reflection.

Crosstalk: In an optical transmission system, leakage of optical power from one optical conductor to another. The leakage may occur by frustrated total reflection from inadequate cladding thickness or low absorptive quality. *See* Fiber crosstalk.

Cylindrical Lens: A lens with a cylindrical surface. . . .

Dark Current: The current that flows in a photodetector when there is no radiant energy or light flux incident upon its sensitive surface (i.e., total darkness). Dark current generally increases with increased temperature for most photodetectors.

Data Bus: In an optical communication system, an optical waveguide used as a common trunk line to which a number of terminals can be interconnected using optical couplers.

Data Bus Coupler: In an optical communication system, a component that interconnects a number of optical waveguides and provides an inherently bidirectional system by mixing and splitting all signals within the component.

Diffraction: The process by means of which the propagation of radiant waves or light waves are modified as the wave interacts with an object or obstacles. Some of the rays are deviated from their path by diffraction at the objects, whereas other rays remain undeviated by diffraction at the objects. As the objects become small in comparison with the wavelength, the concepts of reflection and refraction become useless and diffraction plays the dominant role in determining the redistribution of the rays following incidence upon the objects. Diffraction results in a deviation of light from the paths and foci prescribed by the rectilinear propagation prescribed by geometrical optics. Thus, even with a very small, distant source, some light, in the form of bright and dark bands, is found within a geometrical shadow because of the diffraction of the light at the edge of the object forming the shadow. Gratings with spacings of the order of the wavelength of the incident light cause diffraction. Such gratings can be ruled grids, spaced spots, or crystal lattice structures.

Diffraction Grating: An array of fine, parallel, equally spaced reflecting or transmitting lines that mutually enhance the effects of diffraction at the edges of each so as to concentrate the diffracted light very close to a few directions characteristic of the spacing of the lines and the wavelength of the diffracted light. If I is the angle of incidence, D the angle of diffraction, S the center-to-center distance between successive rulings, N the order of the spectrum, the wavelength is $L = (S/N) (\sin I + \sin D)$. If there is a large number of narrow, close, equally spaced rulings upon a transparent or reflecting substrate, the grating will be capable of dispersing incident light into its frequency component spectrum.

Diffusion: The scattering of light by reflection or transmission. Diffuse reflection results when light strikes an irregular surface such as a frosted window or the surface of a frosted or coated light bulb. When light is diffused, no definite image is formed.

Dispersion: (1) The process by which rays of light of different wavelength are deviated angularly by different amounts, as, for example, with prisms and diffraction gratings. (2) Phenomena that cause the index of refraction and other optical properties of a medium to vary with wavelength. Dispersion also refers to the frequency dependence of any of several parameters, for example, in the process by which an electromagnetic signal is distorted because the various frequency components of that signal have different propagation characteristics and paths. Thus, the components of a complex radiation are dispersed or separated on the basis of some characteristic. A prism disperses the components of white light by deviating each wavelength a different amount. *See* Chromatic dispersion; Fiber dispersion; Material dispersion; . . . Pulse dispersion. . . .

Diverging Lens: A lens that causes parallel light rays to spread out. One surface of a diverging lens may be concavely spherical and the other plane (planoconcave). Both may be concave (double concave) or one surface may be concave and the other convex (concave-convex, divergent-meniscus). The diverging lens is always thicker at the edge than at the center. *Synonyms:* concave lens; dispersive lens; divergent lens; negative lens. The diverging lens is considered to have a negative focal length.

Dopant: A material mixed, fused, amalgamated, crystallized, or otherwise added to another (intrinsic) material in order to achieve desired characteristics of the resulting material. For example, the germanium tetrachloride or titanium tetrachloride used to increase the refractive index of glass for use as an optical-fiber core material, or the gallium or arsenic added to silicon or germanium to produce a doped semiconductor for achieving donor or acceptor, positive or negative material for diode and transistor action.

Double Heterojunction: In a laser diode, two heterojunctions in close proximity, resulting in full carrier and radiation confinement and improved control of recombinations. [See also Heterojunction]

Double Heterojunction Diode: A laser diode that has two different heterojunctions, the difference being primarily in the stepped changes in refractive indices of the material in the vicinity of the p-n junction. The double heterojunction laser diode is widely used for [optical communication]. . . .

Edge-emitting LED: A light-emitting diode with a spectral output that emanates from between the heterogeneous layers (i.e. from an edge), having a higher output intensity and greater coupling efficiency to an optical fiber or integrated optical circuit than the surface-emitting LED, but not as great as the injection laser. Surface-emitting and edge-emitting LEDs provide several milliwatts of power in the spectral range 800–1200 nm at drive currents of 100 to 200 milliamperes; diode lasers at these currents provide tens of milliwatts. *See also* Surface-emitting LED.

Electromagnetic Spectrum: The entire range of wavelengths, extending from the shortest to the longest or conversely, that can be generated physically. This range of electromagnetic wavelengths extends almost from zero to infinity and includes the visible portion of the spectrum known as light. . . .

Electrooptic Effect: The change in the index of refraction of a material when subjected to an electric field. The effect can be used to modulate a light beam in a material, since many properties, such as light conducting velocities, reflection and transmission coefficients at interfaces, acceptance angles, critical angles, and transmission modes, are dependent upon the refractive indices of the media in which the light travels.

End-Fire Coupling: Optical-fiber and integrated optical-circuit (IOC) coupling between two waveguides in which the two waveguides to be coupled are butted up against each other. [It is] a more straightforward, simpler, and more efficient coupling method than evanescent field coupling. Mode pattern matching is required and accomplished by maintaining a unity cross-sectional area aspect ratio, axial alignment, and minimal lateral axial displacement. *See also* Evanescent field coupling.

Evanescent-Field Coupling: Optical-fiber or integrated optical-circuit (IOC) coupling between two waveguides in which the two waveguides to be coupled are held parallel to each other in the coupling region. The evanescent waves on the outside of the waveguide enter the coupled waveguide bringing some of the light energy with it into the coupled waveguide. Close-to-core proximity or fusion is required. . . .

Exitance: *See* Radiant exitance.

Exit Angle: When a light ray emerges from a surface, the angle between the ray and a normal to the surface at the point of emergence. For an optical fiber, the angle between the output ray and the axis of the fiber. . . .

External Optical Modulation: Modulation of a light wave in a medium by application of fields, forces, waves, or other energy forms upon a medium conducting a light beam in such a manner that a characteristic of either the medium, or the beam, or both are modulated in some fashion. External optical modulation can make use of such effects as the electrooptic, acoustooptic, magnetooptic, or absorptive effect.

Fiber: *See* Graded-index fiber; Optical fiber; . . . Single-mode fiber; Step-index fiber.

Fiber Absorption: In an optical fiber, the light-wave power attenuation due to absorption in the fiber core material, a loss usually evaluated by measuring the power emerging at the end of successively shortened known lengths of the fiber.

Fiber Buffer: The material surrounding and immediately adjacent to an optical fiber that provides mechanical isolation and protection. Buffers are generally softer materials than jackets.

Fiber Cladding: A light-conducting material that surrounds the core of an optical fiber and that has a lower refractive index than the core material.

Fiber Core: The central light-conducting portion of an optical fiber. The core has a higher refractive index than the cladding that surrounds it.

Fiber Core Diameter: In an optical fiber, the diameter of the higher refractive index medium that is the primary transmission medium for the fiber.

Fiber Crosstalk: In an optical fiber, exchange of lightwave energy between a core and the cladding, the cladding and the ambient surrounding, or between differently indexed layers. Fiber crosstalk is usually undesirable, since differences in path length and propagation time can result in dispersion, reducing transmission distances. Thus, attenuation is deliberately introduced into the cladding by making it lossy.

Fiber-Detector Coupling: In fiber-optic transmission systems, the transfer of optical signal power from an optical fiber to a detector for conversion to an electrical signal. Many optical-fiber detectors have an optical-fiber pigtail for connection by means of a splice or a connector to a transmission fiber.

Fiber Diameter: The diameter of an optical fiber, normally inclusive of the core, the cladding if step-indexed, and any adherent coating not normally removed when making a connection, such as by a butted or tangential connection.

Fiber Dispersion: The lengthening of the width of an electromagnetic-energy pulse as it travels along a fiber; caused by material dispersion due to the frequency dependence of the refractive index; [by] modal dispersion, [due to] different group velocities of the different modes, and [by] waveguide dispersion due to frequency dependence of the propagation constant of that mode.

Fiber-Optic Cable: Optical fibers incorporated into an assembly of materials that provides tensile strength, external protection, and handling properties comparable to those of equivalent-diameter [electrical] cables. Fiber-optic cables (light guides) are a direct replacement for conventional coaxial cables and wire pairs.

Fiber-Optic Communications (FOC):   Communication systems and components in which optical fibers are used to carry signals from point to point.

Fiber-Optic Multiport Coupler: An optical unit, such as a scattering or diffusion solid "chamber" of optical material, that has at least one input and two outputs, or at least two inputs and one output, that can be used to couple various sources to various receivers. The ports are usually optical fibers. If there is only one input and one output port, it is simply a connector.

Fiber-Optic Rod Coupler: A graded-index cylindrically shaped section of optical fiber or rod with a length corresponding to the pitch of the undulations of light waves caused by the graded refractive index, the light beam being injected via fibers at an off-axis end point on the radius, with the undulations of the resulting wave varying periodically from one point to another along the rod and with half-reflection layers at the ¼-pitch point of the undulations providing for coupling between input and output fibers.

Fiber Optics: The technology of guidance of optical power, including rays and waveguide modes of electromagnetic waves along conductors of electromagnetic waves in the visible and near-visible region of the frequency spectrum, specifically when the optical energy is guided to another location through thin transparent strands. [Technology includes] conveying light or images through a particular configuration of glass or plastic fibers. Incoherent optical fibers will transmit light, as a pipe will transmit water, but not an image. [Aligned bundles of] optical fibers can transmit [a mosaic] image through perfectly aligned, small . . . clad optical fibers. Specialty fiber optics combine coherent and incoherent aspects. . . .

Fiber-Optic Splice: A nonseparable junction joining one optical conductor to another.

Fiber-Optic Transmission System: A transmission system utilizing small-diameter transparent fibers through which light is transmitted. Information is transferred by modulating the transmitted light. These modulated signals are detected by light-sensitive devices (i.e., photodetectors). *See* Laser fiber-optic transmission system.

Fiber-Optic Waveguide: A relatively long thin strand of transparent substance, usually glass, capable of conducting an electromagnetic wave of optical wavelength (visible or near-visible region of the frequency spectrum) with some ability to confine longitudinally directed, or near-longitudinally directed light waves to its interior by means of internal reflection. The fiber-optic waveguide may be homogeneous or radially inhomogeneous with step or graded changes in its index of refraction, the indices being lower at the outer regions, the core thus being of increased index of refraction.

Filter: In an optical system, a device with the desired characteristics of selective transmittance and optical homogeneity, used to modify the spectral composition of radiant light flux. A filter is usually of special glass, gelatin, or plastic optical parts with plane parallel surfaces that are placed in the path of light through the optical system of an instrument to selectively absorb certain wavelengths of light, reduce glare, or reduce light intensity. Colored, ultraviolet, neutral density, and polarizing filters are in common use. Filters may be separate elements or integral devices mounted so that they can be placed in or out of position in a system as desired. . . .

Footcandle: A unit of illumination equal to 1 lumen incident per square foot. It is the illuminance [on] a surface placed 1 foot from a light source having a luminous intensity of 1 . . . candela. [Instead of footcandle, the preferred unit is the "lux," which is 1 lumen per square meter. One footcandle = 10.764 lux.]

Fusion Splicing: In optical transmission systems using solid media, the joining together of two media by butting the media together, forming an interface between them, and then removing the common surfaces so that there be no interface between them when a light wave is propagated from one medium to the other, thus, [ideally] no reflection or refraction can occur at the former interface.

Gain-Bandwidth Product: The product of the [low-frequency] gain of an active device and [the half-power] bandwidth. For an avalanche photodiode, the gain-bandwidth product is the gain times the frequency of measurement when the device is biased for maximum obtainable gain.

Geometric Optics: The optics of light rays that follow mathematically defined paths in passing through optical elements such as lenses and prisms and optical media that refract, reflect, or transmit electromagnetic radiation. The branch of science that treats light propagation in terms of rays, considered as straight or curved lines in homogeneous and nonhomogeneous media.

Graded-Index Fiber: An optical fiber with a variable refractive index that is a function of the radial distance from the fiber axis, the refractive index getting progressively lower away from the axis. This characteristic causes the light rays to be continually refocussed by refraction into the core. As a result, there is a designed continuous change in refractive index between the core and cladding along a fiber diameter. . . .

Graded-Index Profile: The condition of having the refractive index of a material, such as an optical fiber, vary continuously from one value at the core to another at the outer surface.

Heterojunction: In a laser diode, a boundary surface at which a sudden transition occurs in material composition across the boundary, such as a change in the refractive index as well as a change from a positively doped (p) region to a negatively doped (n) region (i.e., a p-n junction) in a semiconductor, or a positively doped region with a rapid change in doping level (i.e., a high concentration gradient of dopant versus distance), and usually at which a change in geometric cross section occurs and across which a voltage or voltage barrier may exist. Heterojunctions provide a controlled degree and direction of radiation confinement, there usually being an equal refractive index step at each heterojunction. *See* Double heterojunction. . . .

Homojunction: In a laser diode, a single junction (i.e., a single region of shift in doping from positive to negative majority carrier regions, or vice versa) and a change in refractive index, at one boundary, hence one energy-level shift, one barrier, and one refractive index shift.

Illuminance: Luminous flux incident per unit area of a surface. Illuminance is expressed in lumens per square meter. *Synonym:* illumination; [luminous incidance]

Incidence Angle: In optics, the angle between the normal to a reflecting or refracting surface and the incident ray.

Incident Ray: A ray of light that falls upon, or strikes, the surface of any object, such as a lens, mirror, prism, this printed page, the things we see, or the human eye. It is said to be incident to the surface.

Index-Matching Materials: Light-conducting materials used in intimate contact to reduce optical power losses by using materials with refractive indices at interfaces that will reduce reflection, increase transmission, avoid scattering, and reduce dispersion.

Infrared Band: The band of electromagnetic wavelengths between the extreme of the visible part of the spectrum, about 750 nm and the shortest microwaves, about 1,000,000 nm. . . .

Injection Laser Diode: A diode operating as a laser producing a monochromatic light modulated by injection of carriers across a *p-n* junction of a semiconductor [having] narrower spatial and wavelength emission characteristics for longer-range higher-data-rate systems than the LEDs, which are more applicable to larger-diameter and larger-numerical aperture fibers for lower-information bandwidths.

Insertion Loss: In light-wave transmission systems, the power lost at the entrance to a waveguide, such as an optical fiber or an integrated optical circuit, due to any and all causes. . . .

Integrated Optical Circuit: A circuit, or group in interconnected circuits, consisting of miniature solid-state optical components, such as light-emitting diodes, optical filters, photodetectors (active and passive), and thin-film optical waveguides on semiconductor or dielectric substrates. *Synonym:* optical integrated circuit.

Integrated Optics: The interconnection of miniature optical components via optical waveguides on transparent dielectric substrates, using optical sources, modulators, detectors, filters, couplers, and other elements incorporated into circuits analagous to integrated electronic circuits for the execution of various communication, switching, and logic functions.

Interference: In light-wave transmission, the systematic reinforcement [and/or] attenuation of two or more light waves when they are superimposed. Interference is an additive process. (The term is applied also to the converse process in which a given wave is split into two or more waves by, for example, reflection and refraction at beam splitters.) The superposition must occur on a systematic basis between two or more waves in order that the electric and magnetic fields of the waves can be additive and produce noticeable effects such as interference patterns. For example, the planes of polarizations should nearly or actually coincide or the wavelengths should [be] nearly or actually . . . the same.

Irradiance: The [radiant flux] per unit area of incident light upon a surface. The radiant flux incident upon a unit area of surface. It can be measured as watts per square meter, as for any form of electromagnetic waves, or as lumens per square meter when visible light is incident upon a surface. The old unit [for luminous incidance] was footcandles. *Synonym:* radiant flux density. . . .

Lambert: A unit of luminance, equal to $10^4/\pi$ [candelas] per square meter. . . .

Laser Diode: A junction diode, consisting of positive and negative carrier regions with a *p-n* transition region (junction) that emits electromagnetic radiation (quanta of energy) at optical wavelengths when injected electrons under forward bias recombine with holes in the vicinity of the junction. In certain materials, such as gallium arsenide, there is a high probability of radiative recombination producing emitted light, rather than heat, at a wavelength suitable for optical waveguides. Some light is reflected by the polished ends and is trapped to stimulate more emission, which further excites, overcoming losses, to produce laser action. *See* Injection laser diode.

Laser Fiber-Optic Transmission System: A system consisting of one or more laser transmitters and associated fiber-optic cables. During normal operation, the laser radiation is limited to the cable. . . .

Laser Line Width: In the operation of a laser, the wavelength range over which most of the laser beam's energy is distributed.

Launch Angle: In an optical fiber . . . the angle between the input radiation vector (i.e., the input light chief ray) and the axis of the fiber. . . . If the ends of the fibers are perpendicular to the axis of the fibers, the launch angle is equal to the angle of incidence when the ray is external . . . and the angle of refraction when initially inside the fiber.

Lens: An optical component made of one or more pieces of a material transparent to the radiation passing through. Having curved surfaces, it is capable of forming an image, either real or virtual, of the object source of the radiation, at least one of the curved surfaces being convex or concave, normally spherical but sometimes aspheric. *See* . . . Converging lens; Diverging lens.

A transparent optical element, usually made from optical glass, having two opposite polished major surfaces of which at least one is convex or concave in shape and usually spherical. The polished major surfaces are shaped so that they serve to change the amount of convergence or divergence of the transmitted rays. . . .

Light: . . . Radiant electromagnetic energy within the limits of human visibility and therefore with wavelengths to which the human retina is responsive. Approximately 380 to 780 nm. . . .

Light-Emitting Diode (LED): A diode that [has applications] similar to [those of] a laser diode, with the same output power level, the same output limiting modulation rate, and the same operational current densities . . . but [has] greater simplicity, tolerance, and ruggedness; and about 10 times the spectral width of [laser diode] radiation.

Light Ray: A line, perpendicular to the wavefront of light waves, indicating their direction of travel and representing the light wave itself.

Light-Wave Communications: That aspect of communications and telecommunications devoted to the development and use of equipment that uses electromagnetic waves in or near the visible region of the spectrum for communication purposes. Light-wave communication equipment includes sources, modulators, transmission media, detectors, converters, integrated optic circuits, and related devices, used for generating and processing light waves. The term "optical communications" is oriented toward the notion of optical equipment, whereas the term "light-wave communications" is oriented toward the signal being processed. *Synonym:* optical commmunications. *See also* Light.

Loose-Tube Splicer: A glass tube with a square hole used to splice two optical fibers; the curved fibers are made to seek the same corner of the square hole, thus holding them in alignment until the index-matching epoxy, already in the tube, cures. This forms an aligned, low-loss butted joint. . . .

Lumen: The SI unit of [luminous] flux corresponding to $1/(4\pi)$ of the total [luminous] flux emitted by a source having an intensity of 1 candela. . . .

Luminance: The . . . luminous intensity [per unit area] emitted by a light source in a given direction by an infinitesimal area of the source. . . . Luminance is usually stated as luminous intensity per unit area (i.e., luminous flux emitted per unit solid angle projected per unit projected area. . . .

Luminous Flux: The quantity that specifies the capacity of the radiant flux . . . to produce . . . visual sensation known as brightness. Luminous flux is radiant flux [weighted by] its luminous [efficacy]. Unless otherwise stated, luminous flux pertains to the standard photooptic observer.

Luminous Intensity: The ratio of the luminous flux emitted by a light source, or an element of the source, in an infinitesimally small cone about the given direction, to the solid angle of that cone, usually stated as luminous flux emitted per unit solid angle.

Lux: A unit of illuminance equal to a lumen incident per square meter of surface. . . .

Magneto-optic Effect: The rotation of the plane of polarization of plane-polarized light waves in a medium brought about when subjecting the medium to a magnetic field (Faraday rotation). The effect can be used to modulate a light beam in a material since many properties, such as conducting velocities, reflection and transmission coefficients at interfaces, acceptance angles, critical angles, and transmission modes, are dependent upon the direction of propagation at interfaces in the media in which the light travels. . . . The magnetic field is in the direction of propagation of the light wave. *Synonym:* Faraday effect.

Material Dispersion: [Bandwidth limitation due to variable material property in] an optical transmission media used in optical waveguides, such as the variation in the refractive index of a medium as a function of wavelength, optical fibers, slab dielectric waveguides, and integrated optical circuits. Material dispersion contributes to group-delay distortion, along with waveguide-delay distortion and multimode group-delay spread. The part of the total dispersion of an electromagnetic pulse in a waveguide caused by the changes in properties of the material with which the waveguide such as an optical fiber is made, due to changes in frequency. As wavelength increases . . . material dispersion decreases; at [short wavelengths] the rapid interactions of the electromagnetic field with the waveguide material (optical fiber) renders the refractive index even more dependent upon wavelength.

Maximum Acceptance Angle: The maximum angle between the longitudinal axis of an optical transmission medium, such as an optical fiber or a deposited optical film, and the normal to the wavefront [at which propagation can take place] (i.e., the direction of the entering light ray), in order that there be total internal reflection of the portion of incident light that is transmitted through the fiber interface (i.e., the [input angle for which] the angle between the transmitted ray and the normal to the inside surface of the cladding is greater than the critical angle). The [sine of the] maximum acceptance angle is given by the square root of the difference of the squares of the indices of refraction of the fiber core glass and the cladding. The square root of the difference of the squares is called the numerical aperture (NA).

Microbending Loss: In an optical fiber, the loss or attenuation in signal power caused by small bends, kinks, or abrupt discontinuities in direction of the fibers, usually caused by fiber cabling or by wrapping fibers on drums. Microbending losses usually result from a coupling of guided modes among themselves and among the radiation modes.

Mirror: A flat surface optically ground and polished on a reflecting material, or a transparent material that is coated to make it reflecting, used for reflecting light. A beam-splitting mirror has a lightly deposited metallic coating that transmits a portion of the incident light and reflects the remainder. A smooth highly polished plane or curved surface for reflecting light. Usually, a thin coating of silver or aluminum on glass constitutes the actual reflecting surface. When this surface is applied to the front face of the glass, the mirror is a front-surface mirror. . . .

Monochromatic: Pertaining to a composition of one color. Purely monochromatic light has all its energy confined to one . . . wavelength.

Monochromatic Light: Electromagnetic radiation, in the visible or near-visible (light) portion of the spectrum, that has only one frequency or wavelength.

Monochromatic Radiation: Electromagnetic radiation that has one . . . wavelength. . . .

Multichannel Cable: In optical-fiber systems, two or more cables combined in a single jacket, harness, strength member, cover, or other unitizing element.

Near Infrared: Pertaining to electromagnetic wavelengths from 750 to 3000 nm.

Noise-equivalent Power (NEP): In optics, the [bandwidth-normalized (per square root of bandwidth)] value of optical power required to produce unity rms signal-to-noise ratio. NEP is a common parameter in specifying detector performance [in watts per root hertz]. NEP is useful for comparison only if modulation frequency, bandwidth, detector, area, and temperature are specified. NEP is indicated in watts [per root hertz].

Nonreflective Star Coupler: An optical-fiber coupling device that enables signals in one or more fibers to be transmitted to one or more other fibers by entering the input signal fibers into an optical-fiber volume without an internal reflecting surface so that the diffused signals pass directly to the output fibers on the opposite side of the fiber volume for conduction away in one or more of the output fibers. The optical-fiber volume is a shaped piece of the optical-fiber material to achieve transmission of two or more inputs to two or more outputs. *See also* Reflective star coupler; Tee coupler.

Numerical Aperture (NA):   A measure of the light-accepting property of an optical fiber (e.g., glass), given by NA = square root of the difference of the squares of the indices of refraction of the core n (1), and the cladding, $n(2)$. If $n(1)$ is 1.414 (glass) and $n(2)$ is 1.0 (air), the numerical aperture is 1.0, and all incident rays will be trapped. The numerical aperture is a measure of the characteristic of an optic conductor in terms of its acceptance of impinging light. The degree of openness, light-gathering ability, angular acceptance, and acceptance cone are all terms describing this characteristic. It may be necessary to specify that the indices of refraction are for step-index fibers and for graded-index fibers $n(1)$ is the maximum index in the core and $n(2)$ is the minimum index in the cladding. As a number, the NA expresses the light-gathering power of a fiber. It is mathematically equal to the sine of the acceptance angle. . . . The numerical aperture is also equal to the sine of the half-angle of the widest [conical] bundle of rays capable of entering a lens, multiplied by the index of refraction of the medium containing that bundle of rays (i.e., the incident medium).

Optical Attenuator: In an optical-fiber data link or integrated optical circuit, a device used to reduce the intensity (i.e., attenuate the light waves when inserted into an optical waveguide). Three basic forms of optical attenuators have been developed: a fixed optical attenuator, a stepwise variable optical attenuator, and a continuous variable optical attenuator. One form of attenuator uses a filter consisting of a metal film evaporated onto a sheet of glass to obtain the attenuation. The filter might be tilted to avoid reflection back into the input optical fiber or cable. . . .

Optical Directional Coupler: A device used in optical-fiber communication systems, such as CATV and data links for optical-fiber measurements, to combine or split optical signals at desired ratios by insertion into a transmission line, for example, a three-port or four-port unit with precise connectors at each port to enable inputs to be coupled together and transmitted via multiple outputs.

Optical Fiber: A single discrete optical transmission element usually consisting of a fiber core and a fiber cladding. As a light-guidance system (dielectric waveguide) that is usually cylindrical in shape, it consists either of a cylinder of transparent dielectric material of given refractive index whose walls are in contact with a second dielectric material of a lower refractive index, or of a cylinder whose core has a refractive index that gets progressively lower away from the center. The length of a fiber is usually much greater than its diameter. The fiber relies upon internal reflection to transmit light along its axial length. Light enters one end of the fiber and emerges from the opposite end with losses dependent upon length, absorption scattering, and other factors. . . .

Optical-Fiber Bundle: Many optical fibers in a single protective sheath or jacket. The jacket is usually polyvinyl chloride (PVC). The number of fibers might range from a few to several hundred, depending on the application and the characteristics of the fibers.

Optical-Fiber Coating: A protective material bonded to an optical fiber, over the cladding if any, to preserve fiber strength and inhibit cabling losses, by providing protection against mechanical damage, protection against moisture and debilitating environments, compatibility with fiber and cable manufacture, and compatibility with the jacketing process. Coatings include fluorpolymers, Teflon, Kynar, polyurethane, and many others. . . .

Optical Repeater: An optical/optical, optical/electrical, or electrical/optical signal amplification and processing device.

Optical Surface: In an optical system, a reflecting or refracting surface of an optical element, or any other identified geometric surface in the system. Normally, optical surfaces occur at surfaces of discontinuity (abrupt changes) of fractive indices, absorptive qualities, transmissivity, vitrification, or other optical quality or characteristic.

Optical Transmitter: A source of light capable of being modulated and coupled to a transmission medium such as an optical fiber or an integrated optical circuit.

Optics: That branch of physical science concerned with the nature and properties of electromagnetic radiation and with the phenomena of vision. . . .

Optoelectronic Device: (1) A device [that emits or is] responsive to electromagnetic radiation in the visible, infrared, or ultraviolet spectral regions of the frequency spectrum [converting electric signals to optical, or vice versa] . . . or utilizes such electromagnetic radiation for its internal operation. The wavelengths handled by these devices range from approximately 300 to 30,000 nm. (2) Electronic devices associated with light, serving as sources, conductors, or detectors. . . .

Photodetector: A device capable of [detecting or] extracting the information from an optical carrier, (i.e., a thermal detector or a photon detector, the latter being used for communications more than the former. . . .

Photodetector Responsivity: The ratio of the rms value of the output current or voltage of a photodetector to the rms value of the . . . optical power input. In most cases, detectors are linear in the sense that the responsivity is independent of the intensity of the incident radiation. Thus, the detector response in amps or volts is proportional to incident optical power, watts . . . .

Photon: A quantum of electromagnetic energy. The energy of a photon is [h /c/λ), where h is Planck's constant, c is the speed of light, and λ is the wavelength]. . . .

PIN Diode: A junction diode doped in the forward direction positive, intrinsic, and negative, in that order. PIN diodes are used as photodetectors in fiber and integrated optical circuits.

Prism: A transparent body with at least two polished plane faces inclined with respect to each other, from which light is reflected or through which light is refracted. When light is refracted by a prism whose refractive index exceeds that of the surrounding medium, it is deviated or bent toward the thicker part of the prism. . . .

Pulse Dispersion: A separation or spreading of input optical signals along the length of a transmission line, such as an optical fiber. This limits the useful transmission bandwidth of the fiber. It is expressed in time and distance as nanoseconds per kilometer. Three basic mechanisms for dispersion are the material effect, the waveguide effect, and the multimode effect. Specific causes include surface roughness, presence of scattering centers, bends in the guiding structure, deformation of the guide, and inhomogeneities of the guiding medium. *Synonym:* pulse spreading.

Radiance: The radiant intensity of electromagnetic radiation per unit projected area of a source or other area (i.e., it is the radiant power of electromagnetic radiation per unit solid angle and per unit surface area normal to the direction considered). The surface may be that of a source detector, or it may be any other real or virtual surface intersecting the flux. The unit of measure is watts/steradian-square meter. . . . *Synonym:* emittance. . . .

Radiant Exitance: The radiant power emitted [in all directions] by a unit area of source.

Radiant Flux: The time rate of flow of radiant energy. The units are watts, or joules/second. The radiant energy crossing or striking a surface per unit time, usually measured in watts.

Radiant Intensity: The radiant power per unit solid angle in the direction considered (i.e., the time rate of transfer of radiant energy per unit solid angle, or the flux radiated per unit solid angle about a specified direction). The unit of measure is watts/steradian or joules/(steradian-second). . . .

Radiant Power: The time rate of flow of electromagnetic energy. The unit is watts or joules/second. [The preferred term is "radiant flux."]

Radiant Transmittance: The ratio of the radiant flux transmitted by an object to the incident radiant flux.

Radiation: The electromagnetic waves or photons emitted from a source. . . .

Radiometry: The science devoted to the measurement of radiated electromagnetic [flux or flux per unit area]. In light-wave communications and the use of optical fibers, primary concern is devoted to radiometry rather than photometry. [Photometry deals with measurement of optical radiation in the visible spectrum (i.e., luminous flux).]

Reflectance: The ratio of the reflected flux to the incident flux. This term is applied to radiant and to luminous flux. Unless qualified, reflectance applies to specular, or regular, reflection. . . .

Reflection: When electromagnetic waves, more appropriately light rays, strike a smooth polished surface, their return or bending back into the medium from whence they came. Specular or regular reflection from a polished surface, such as a mirror, will return a major portion of the light in a definite direction lying in the plane of the incident ray and the normal after specular reflection. Light can be made to form a sharp image of the original source. Diffuse reflection occurs when the surface is rough and the reflected light is scattered from each point in the surface. These diffuse rays cannot be made to form an image of the original source, only of the diffusely reflecting surface itself. *See* Total internal reflection; Snell's law.

Reflection Angle: When a ray of electromagnetic radiation strikes a surface, and is reflected in whole or in part by the surface, the angle between the normal to the reflecting surface and the reflected ray. *See* Critical angle.

Reflective Star Coupler: An optical-fiber coupling device that enables signals in one or more fibers to be transmitted to one or more other fibers by [injecting] the signals into one side of an optical cylinder, fiber, or other piece of material, with a reflecting back surface so as to reflect the diffused signals back to the output ports on the same side of the material, for conduction away in one or more fibers. *See also* Tee coupler. . . .

Refraction: The bending of oblique (nonnormal) incident electromagnetic waves or rays as they pass from a medium of one index of refraction into a medium of a different index of refraction, coupled with the changing of the velocity of propagation of the electromagnetic waves when passing from one medium to another with different indices of refraction. The waves or rays are usually changed in direction (i.e., bent) crossing the media interface. *See* Refractive index; Snell's law.

Refraction Angle: When an electromagnetic wave strikes a surface and is wholly or partially transmitted into the new medium, of which the struck surface is the boundary, the acute angle between the normal to the refracting surface at the point of incidence, and the refracted ray.

Refractive Index: (1) The ratio of the velocity of light in a vacuum to the velocity of light in the medium whose index of refraction is desired, for example, $n = [1.6]$ for certain kinds of glass. (2) [Relative refractive index is the] ratio of the sines of the angle of incidence and the angle of refraction when light passes from one medium to another. The index between two media is the relative index, while the index when the first medium is a vacuum is the absolute index of the second medium. The index of refraction expressed in tables is the absolute index, that is, vacuum to substance at a certain temperature, with light of a certain wavelength. Examples, vacuum, 1.000; air, 1.000292; water, 1.333; ordinary crown glass, 1.516. Since the index of air is very close to that of vacuum, the two are often used interchangeably. *Synonyms:* absolute refractive index; index of refraction.

Single-Mode Fiber: A fiber waveguide that supports the propagation of only one mode. The single-mode fiber is usually a low-loss optical waveguide with a very small core. . . . It requires a laser source for the input signals because of the very small entrance aperture (acceptance cone). The small core radius approaches the wavelength of the source; consequently, only a single mode is propagated. [Mode is, in simple terms, the path of an optic ray.]

Skew Ray: In an optical fiber, a light ray that never intersects the axis of the fiber while being internally reflected. The skew ray is at an angle to the fiber axis. If the fiber waveguide is straight, a skew ray traverses a helical path along the fiber, not crossing the fiber axis. A skew ray is not confined to the meridian plane. The skew ray is not a meridional ray.

Snell's Law: When electromagnetic waves such as light pass from a given medium to a denser medium, its path is deviated toward the normal; when passing into a less dense medium, their path is deviated away from the normal. Snell's law, often called the law of refraction, defines this phenomenon by describing the relation between the angle of incidence and the angle of refraction as $\sin i / \sin r = n(r)/n(i)$, where i is the angle of incidence, r is the angle of refraction, $n(r)$ is the refractive index of the medium containing the refracted ray, and $n(i)$ is the refractive index containing the incident ray. Stated in another way, both laws, that of reflection and of refraction, are attributed to Snell: namely, when the incident ray, the normal to the surface at the point of incidence of the ray on the surface, the reflected ray, and the refracted ray all lie in a single plane. The angle between the incident ray and the normal is equal in magnitude to the angle between the reflected ray and the normal. The ratio of the sine of the angle between the normal and the incident ray to the sine of the angle between the normal and the refracted ray is a constant. *See also* Refraction.

Solid-State Laser: A laser whose active medium is a solid material such as glass, crystal, or semiconductor material rather than gas or liquid.

Source-Coupler Loss: In an optical data link, optical communication system, or optical-fiber system, the loss, usually expressed in dB, between the light source and the device or material that couples the light source energy from the source to the fiber cable.

Source-Fiber Coupling: In fiber-optic transmission systems, the transfer of optical signal power emitted by a light source into an optical fiber, such coupling being dependent upon many factors, including geometry and fiber characteristics. Many optical-fiber sources have an optical-fiber pigtail for connection by means of a splice or a connector to a transmission fiber.

Source-to-Fiber Loss: In an optical fiber, signal power loss caused by the distance of separation between a signal source and the conducting fiber.

Spectral Bandwidth: The wavelength interval in which a radiated spectral quantity is a specified fraction of its maximum value. The fraction is usually taken as 0.50 of the maximum power level. . . . If the electromagnetic radiation is light, it is the radiant intensity half-power points that are used.

Spontaneous Emission: In a laser, the emission of light that does not bear an amplitude, phase, or time relationship with an applied signal and is therefore a random noise-like form of radiation.

Step-Index Fiber: A fiber in which there is an abrupt change in refractive index between the core and cladding along a fiber diameter, with the core refractive index higher than the cladding refractive index. These may be more than one layer, each layer with a different refractive index that is uniform throughout the layer, with decreasing indices in the outside layer.

Step-Index Profile: The condition of having the refractive index of a material, such as an optical fiber, change abruptly from one value to another at the core–cladding interface, or at other interfaces if several layers are present.

Steradian: The unit of solid angular measure; [for a solid angle with its apex at the center of a sphere, the measure in steradians is] the subtended surface area of [the] sphere divided by the square of the sphere radius. There are 4π steradians in a sphere. The solid angle subtended by a cone of half-angle A is $2\pi(1 - \cos A)$ steradians.

Surface-Emitting LED: A light-emitting diode with a spectral output that emanates from the surface of the layers, having a lower output intensity and lower coupling efficiency to an optical fiber or integrated optical circuit than the edge-emitting LED and the injection laser. Surface-emitting LEDs provide several milliwatts of power in the spectral range 800 to 1200 nm at drive currents of 100 to 200 milliamperes; diode lasers at these currents provide tens of milliwatts. *Synonyms:* front-emitting LED; Burrus LED. *See also* Edge-emitting LED.

Tee Coupler: In an optical fiber, a reflective surface placed inside the fiber, at 45 degrees to the direction of wave propagation, allowing a part of the signal power to be reflected from one side of the surface out of the fiber at right angles in one direction, and an input signal from the other side of the fiber to be reflected from the other side of the 45-degree reflective surface so as to propagate in the fiber, longitudinally, in the same direction as the original signal to which the input signal is being added and the output signal is being taken. Two tee-couplers can be combined in a single unit for input and output of signals in both directions of propagation. In addition to an optical component used to interconnect a number of terminals through optical waveguides by using partial reflections at dielectric interfaces or metallic surfaces, coupling can be accomplished simply by splitting the waveguide bundle so that fractions can diverge in different directions. See also Reflective star coupler; Nonreflective star coupler.

Thick Lens: A lens whose axial thickness is so large that the principal points and the optical center cannot be considered as coinciding at a single point on its optical axis.

Thin-Film Optical Modulator: A device made of multilayered films of material of different optical characteristics capable of modulating transmitted light by using electro-optic, electroacoustic, or magneto-optic effects to obtain signal modulation. Thin-film optical modulators are used as component parts of integrated optical circuits.

Thin-Film Optical Multiplexer: A multiplexer consisting of layered optical materials that make use of electro-optic, electroacoustic, or magneto-optic effects to accomplish the multiplexing. Thin-film optical multiplexers may be component parts of integrated optical circuits.

Thin-Film Optical Switch: A switching device for performing logic operations using light waves in thin films, usually supporting only one propagation mode, making use of electro-optic, electroacoustic, or magneto-optic effects to perform switching functions, such as are performed by semiconductor gates (*AND, OR,* negation). Thin-film optical switches may be component parts of integrated optical circuits.

Thin-Film Optical Waveguide: An optical waveguide consisting of thin layers of differing refractive indices, the lower indexed material on the outside or as a substrate, usually for supporting a single electromagnetic wave propagation mode with laser sources. The thin-film waveguide lasers, modulators, switches, directional couplers, filters, and related components need to be coupled from their integrated optical circuits to the optical waveguide transmission media, such as optical fibers and slab dielectric waveguides.

Thin Lens: A lens whose axial thickness is sufficiently small that the principal points, the optical center, and the vertices of its two surfaces can be considered as coinciding at the same point on its optical axis.

Total Internal Reflection: The reflection that occurs within a substance because the angle of incidence of light striking the boundary surface is in excess of the critical angle. *See also* Critical angle.

Transmittance: The ratio of the flux that is transmitted through an object, to the incident radiant or luminous flux. Unless qualified, the term is applied to regular (i.e., specular) transmission. . . .

Waveguide Delay Distortion: In an optical waveguide, such as an optical fiber, dielectric slab waveguide, or an integrated optical circuit, the distortion in received signal caused by the differences in propagation time for each wavelength (i.e., the delay versus wavelength effect for each propagating mode), causing a spreading of the total received signal at the detector. Waveguide delay distortion contributes to group-delay distortion, along with material dispersion and multimode group-delay spread.

Waveguide Dispersion: The part of the total dispersion attributable to the dimensions of the waveguide since they are critical for modes allowed and not allowed to propagate, such that waveguide dispersion increases as frequency decreases, due to these dimensions and their variation along the length of the guide.

Wavelength: The length of a wave measured from any point on one wave to the corresponding point on the next wave, such as from crest to crest. Wavelength determines the nature of the various forms of radiant energy that comprise the electromagnetic spectrum; for example, it determines the color of light. . . . [Wavelengths of particular sources are usually given as their wavelengths in free space.]

Wavelength Division Multiplex (WDM): In optical communication systems, the multiplexing of light waves in a single medium such as a bundle of fibers, such that each of the waves is of a different wavelength and is modulated separately before insertion into the medium. Usually, several sources are used, such as a laser, or a dispersed white source, each having a distinct center wavelength. WDM is the same as frequency-division multiplexing (FDM) applied to other than visible light frequencies of the electromagnetic spectrum.

Appendix A

FURTHER READING

MAGAZINES

EDN
221 Columbus Avenue, Boston, Massachusetts 02116
Electronic Design
50 Essex St., Rochelle Park, New Jersey 07662
Electro-Optical Systems Design
222 West Adams St., Chicago, Illinois 60606
IEEE Spectrum
345 East 47th Street, New York, New York 10017
Laser Focus
1001 Watertown St., Newton, Massachusetts 02165
Proceedings of the IEEE
345 East 47th Street, New York, New York 10017

BOOKS (In alphabetical order)

Fiber and Integrated Optics, edited by D. B. Ostrowsky, Plenum Press, New York 1979.

Fiber Optics—Advances in Research and Development, Plenum Press, New York 1979.

Fiber Optics Handbook and Buyer's Guide, Information Gatekeepers, Brookline, Mass.

Fiber Optics in Communications Systems, by Glenn R. Elion and Herbert A. Elion, Marcel Dekker, Inc., New York, 1978.

Fundamentals of Optical Fiber Communications, by M. K. Barnowski, Academic Press, Inc., New York, 1976.

Handbook of Fiber Optics: Theory and Applications, edited by Helmut F. Wolf, Garland Publishing, New York, N.Y., 1980.

Optical Fibre Communications, by C. P. Sandbank, John Wiley & Sons, Inc., New York, 1980.

Optical Fibre Communications: Devices, Circuits and Systems, edited by M. J. Howes and D. V. Morgan, John Wiley & Sons Ltd., Chichester, England, 1980.

Optical Fibers for Transmission, by J. E. Midwinter, John Wiley & Sons, Inc., New York, 1979.

Optical Fiber Telecommunications, edited by S. E. Miller and A. G. Chynoweth, Academic Press, Inc., New York, 1979.

Appendix B

ADDRESSES

**OPTICAL FIBER
MANUFACTURERS**

Belden Corporation
2000 South Batavia Avenue
Geneva, Illinois 60134

Corning
Telecommunication Products
    Department
Corning Glass Works
Corning, New York 14830

Du Pont
E. I. Du Pont de Nemours &
    Company
Wilmington, Delaware 19898

General Cable Company
Communications Products
    Operation
P.O. Box 700
1 Woodbridge Center
Woodbridge, New Jersey 07095

ITT
Electro-Optical Products Division
7635 Plantation Road
Roanoke, Virginia 24019

**CONNECTOR
MANUFACTURERS**

AMP Incorporated
Harrisburg, Pennsylvania 17105

Amphenol North America
Bunker Ramo Corporation
33 East Franklin Street
Danbury, Connecticut 06810

ITT
Cannon Electric Division
P.O. Box 929
Santa Ana, California 92702

Sealectro Corporation
Mamaroneck, New York 10543

TRW Cinch Connectors
1500 Morse Avenue
Elk Grove Village, Illinois 60007

Trompeter Electronics Inc.
8936 Comanche Avenue
Chatsworth, California 91311

TEST EQUIPMENT MANUFACTURERS

Bowmar
531 Main Street
Acton, Massachusetts 01720

Fiberguide Instruments
1101B State Road
Princeton, New Jersey 08540

Hamamatsu Corporation
420 South Avenue
Middlesex, New Jersey 08846

Photodyne Inc.
5356 Sterling Center Drive
Westlake Village, California 91361

Siecor Optical Cables, Inc.
P.O. Box 489
Hickory, North Carolina 28601

Times Fiber Communications, Inc.
358 Hall Avenue
Wallingford, Connecticut 06492

SYSTEM/LINK MANUFACTURERS

AEG-Telefunken
P.O. Box 1109
D-71 Heilbronn, West Germany

Burr-Brown
P.O. Box 11400
International Airport Industrial
Park
Tucson, Arizona 85734

Fujitsu Limited
6-1, Marunouchi 2-chome
Chiyoda-ku
Tokyo 100, Japan
and
680 Fifth Avenue
New York, New York 10019

Harris Corporation
P.O. Box 37
Melbourne, Florida 32901

Hewlett-Packard
Optoelectronics Division
640 Page Mill Road
Palo Alto, California 94304

ITT Telecommunications
Corporation
Transmission Division
2912 Wake Forrest Road
Raleigh, North Carolina 27611

LeCroy
Fiberoptic Systems Division
1234 East Joppa Road
Towson, Maryland 21204

Lignes Telegraphiques et Telep
LTT 1 rue Charles Bourseul
78707 Conflans-Ste. Monorine
France

Math Associates, Inc.
376 Great Neck Road
Great Neck, New York 11021

NEC America, Inc.
Radio & Transmission Division
2990 Telestar Ct., Suite 212
Falls Church, Virginia 22042

Optical Information Systems
Exxon Enterprises Inc.
350 Executive Boulevard
Elmsford, New York 10523

RCA
Electro-Optics
Lancaster, Pennsylvania

RCA
Solid-State Division
Route 202
Somerville, New Jersey 08876

T&B/Ansley Corporation
3208 Humboldt Street
Los Angeles, California 90031

LED/LASERS/PHOTODIODES

General Optronics
3005 Hadley Road
South Plainfield, New Jersey 07080

Laser Diode Laboratories, Inc.
1130 Somerset Street
New Brunswick, New Jersey 08901

Motorola Semiconductor
5005 East McDowell Road
Phoenix, Arizona 85008

RCA
Solid-State Division
Route 202
Somerville, New Jersey 08876

Texas Instruments Incorporated
Dallas, Texas 75265

INDEX

INDEX